earth

by the same authors
The End of Affluence
The Golden Door (with D.L. Bilderback)
Extinction
Ecoscience (with John P. Holdren)

BY PAUL EHRLICH
The Process of Evolution (with Richard Holm)
The Population Bomb
How To Be A Survivor (with Richard Harriman)
Ark II (with Dennis Pirages)
The Machinery of Nature
The Science of Ecology (with Jonathan Roughgarden)

The photograph on p. 1 is of Planet
Earth as seen from the moon.
It was taken as the Apollo 8
expedition orbited the moon
in December 1968.

earth

by

ANNE H. EHRLICH AND PAUL R. EHRLICH

Franklin Watts   New York   1987

To LuEsther, Sis and Peck,
whose friendship has meant so much to us

Copyright © 1987 Anne H. Ehrlich and Paul R. Ehrlich
Picture Editor: Isobel Hinshelwood
All rights reserved.
First published in England in 1987 by Methuen London Ltd.

First published in the United States in 1987 by Franklin Watts, Inc.
387 Park Avenue South, New York, NY 10016
Library of Congress Catalog Card Number: 86–50870
ISBN 0–531–15036–4

Printed in Great Britain

Contents

List of Illustrations

Acknowledgements

The authors gratefully acknowledge the following individuals who reviewed and commented on part or all of the manuscript: Herman Daly (Department of Economics, Louisiana State University), John Harte (Energy and Resources Group, University of Califonia, Berkeley), Mary Ellen Harte (Paleontology Museum, University of California, Berkeley), Sarah Hoffman (College of Oceanography, Oregon State University), Cheryl E. Holdren (Rocky Mountain Biological Laboratory and Center for Conservation Biology, Stanford University), John P. Holdren (Energy and Resources Group, University of California, Berkeley), Richard W. Holm (Department of Biological Sciences, Stanford University), Stephen Schneider (National Center for Atmospheric Research) and Ward B. Watt (Department of Biological Sciences, Stanford University). Their suggestions have undoubtedly greatly improved the book.

In addition, the authors and the photo editor appreciate the help of Tony Angell, Judith Kunofsky and Pamela Matson in finding and obtaining photos and illustrations, and of Miriam Rothschild, who provided a photo and information about her father. As always, this book might never have been completed without the aid and support of Claire Schoens and other staff members of Falconer Library at Stanford, and of Steve Masley and Pat Browne, who operate the copying machine.

Finally, we are all grateful to the production team of Thames Television, Richard Broad, Mike Dormer and Anthony Lee, to Jack Frogell and Carl Fuss of Thames Television's Visual Services for processing reams of photographic material, and to Julia Brown of Thames Methuen for additional picture advice.

PART ONE

An Island of Life

2. **Early Earth.** An artist's impression of the volcanically violent young Earth, as the surface crust was beginning to cool and oceans were condensing from water vapour in the atmosphere.

CHAPTER ONE

Only a Little Planet

From a galactic perspective, Earth is not terribly impressive. The planet, one of the smaller ones in the solar system, is only 25,000 miles around the equator; we personally have flown light aircraft almost ten times that far. Earth's diameter is less than a tenth, and its mass less than a three-hundredth, that of the system's largest planet, Jupiter. Its diameter is less than a thousandth, and its mass less than a three-hundred-thousandth, that of the sun (itself only a medium-sized star, among quintillions of stars arrayed in billions of galaxies). Earth is, to be frank, an inconceivably insignificant mote in the universe. But it's all we have.

Astronomical observations and planetary exploration have not turned up any other suitable habitat for *Homo sapiens*. Although some scientists conjecture that there may be planets in other star systems that could support the kind of life we know, even the existence of planets in such systems remains, with one possible exception, just an educated guess. What is not conjectural is that, at least for the foreseeable future, human beings will be confined to the solar system; the vast majority of people will spin out their lives on the only planet in the universe that we *know* can support human life.

Earth itself has not always had that capacity. For half or more of the 4.6 billion or so years since it condensed from a whirling cloud of gases and dust, along with the sun and its sister planets, Earth was quite an inhospitable place by human standards. The atmosphere was nearly devoid of oxygen, and the emerging continents were barren of plant or animal life and bathed in poisonous ultra-violet radiation.

During its earliest ages, Earth would have seemed quite hostile to life in most of its present forms. While still cooling, the planet was bombarded for several hundred million years with meteors, including thousands that were 10 kilometres or more in diameter – the same bombardment that pocked the moon with gigantic craters. On Earth, the prolonged assault from space probably contributed to violent volcanic activity, which, together with high levels of radioactivity, may have kept the surface molten and possibly helped generate the atmosphere. The accumulation of meteorites also may have increased the planet's mass by 20 per cent or more.

Not until the bombardment abated could the planet cool sufficiently to form a thin surface crust and allow the oceans to condense out of the abundant water vapour in the atmosphere. Soon after forming, however, possibly well over four billion years ago, the oceans appear to have contained all the necessary ingredients for sustaining life. Besides water (hydrogen and oxygen), all the major and minor elements, from nitrogen, carbon, phosphorus and sulphur to iron, silicon and cobalt, seem to have been present virtually from the very start.

Very early in Earth's history, perhaps while the oceans were still forming, pieces of the planet's newly formed solid outer shell – called tectonic plates – began to shift around on the more plastic layer beneath them, related to the formation and gradual spreading of the sea floors. This process continues today. New sea floor is continually created by molten material rising from the planet's plastic mantle along ridges at the bottom of the oceans. At the edges of the plates, old sea floor material is forced beneath the surface and swallowed again – subducted – deep into the interior, causing earthquakes and volcanic activity on the surface.

Scientists have long believed that life began in the shallow seas of the young Earth. One early theory held that the oceans then consisted of a thin 'organic soup', containing complex molecules whose basic ingredient was carbon. Carbon is a most unusual element. It is capable of combining with a wide variety of other elements, and a single carbon atom can combine with four other atoms (of carbon or other elements) at the same time, which enables it to form a staggering number of chemical compounds. Because of its remarkable properties, carbon became the chemical basis of life as we know it.

The carbon compounds that formed in the early oceans of Earth are thought to have been a sample of the simplest kinds of molecules that now characterise living systems (organic molecules). But these primitive organic molecules were not synthesised (assembled) by plants, animals or micro-organisms. It has been speculated that they might have been products of physical events such as the passage of lightning through the atmosphere or chemical reactions taking place on the surface of clays, with the ultra-violet rays in sunlight providing the energy for their synthesis. Under appropriate conditions, these organic compounds apparently form quite readily. Not only have they been created many times in laboratories, but they have been found in the remains of meteorites and in samples of material from space.

Another possibility for the origin of organic molecules on Earth, recently proposed, is that they might have been produced abundantly at the bottom of the oceans in geothermal 'vents' along the mid-ocean

ridges where new sea floor was (and still is) constantly being formed.
In these submarine hot springs, called 'black smokers', mineral-rich
superheated fluids and gases escape from beneath the sea floor and
mix with seawater.

3. **A black smoker.** At depths of 700 to 4,000 metres beneath the ocean's surface, peculiar ecosystems associated with submarine vents – black smokers – have been found. These systems are supported mainly by energy wrested from inorganic chemical reactions by several different types of bacteria and include fascinating faunas of tube worms, crabs, mussels, clams and weird fishes, some of which can be seen at the right in this photograph. Part of the deep-sea diving apparatus is visible in the foreground. Some of the tube worms in these systems have mutualistic relationships with bacteria living inside their bodies; the other animals simply graze the free-living bacteria or prey on other animals that do. The vents are not widespread today, and each one seems to be active for only a few years or decades.

In Earth's early history, subsea volcanic activity was doubtless much
more intense and widespread than it is today and the oceans were
probably shallower; thus more of the ocean water could have been
exposed to the venting process. The physical and chemical conditions
in and around the submarine hydrothermal vents appear to have been
suitable for the frequent formation of organic molecules and for the
further development of these into simple proteins and other more
complex structures, including some that crudely resemble the cells of
living organisms.

It was once thought that this state of affairs, in which organic
molecules formed and disintegrated, mixed and combined, were
cooked and were cooled, went on for several hundred million years
before life arose. Indeed, billions of years might have been needed for
more complex entities to be produced from an 'organic soup' in the
ocean or in tide pools, as was then supposed. But, if the forming,
mixing and recombining of organic chemicals took place in the
pressure-cooker environment of submarine hot springs, as many
scientists now think, the next step might have followed rather quickly
(in geological terms), and there is a fair amount of geological evidence
that it did.

4. Surtsey volcano. The dynamic processes of crustal movement that produced Earth's continents and islands are still going on today. The island of Surtsey, off the coast of Iceland, first emerged from the sea in 1963 as an active volcano; within four years it had attained a height of over 500 feet, and life had begun to occupy it.

However they arose, in ways that can only be guessed at today, some of the more complex organic molecules began to play a part in their own construction and that of other kinds of molecules. The all-important first step in the origin of life, the self-replication of organic molecules, took place. After that it was 'easy', given that vast amounts of time were available. Molecules that could copy themselves became more common than those that could not. By chance, some formed inside complexes that shielded the duplicators from the environment, and these could reproduce themselves even more successfully. The distinction between organisms and environment appeared; self could be separated from non-self; life had originated.

5. Grand Canyon. As land areas are built up through volcanic and tectonic action, they are worn down by processes of erosion – the forces of wind and water exerted over many ages. In the Grand Canyon in Arizona, the oldest layers of rock exposed by the cutting action of the Colorado River are hundreds of millions of years old.

Some evidence now suggests that, because of the continuing production by geophysical processes of organic chemicals in the ocean depths, life may have originated not just once, but numerous times. Exactly when it first happened is not known, and might never be known. Precious little in the way of geological remains has survived intact for more than three billion years – although perhaps it should be considered remarkable that anything has, given Earth's tumultuous history. None the less, some of the oldest sedimentary rocks ever found, about three and a half billion years old, have been found to contain fossils of bacteria similar to some that exist today, suggesting that life began very early in Earth's history.

With the appearance of life, evolution was on its way, with the basic rules under which it now operates already established. Different kinds of self-replicating molecules reproduced themselves at different rates, just as different kinds of organisms do today. This was the beginning of 'natural selection', which can be defined simply as the differential reproduction of genetic 'types' (organisms with different hereditary endowments). Those types that reproduce themselves more often and more successfully tend to replace the less successful ones. Natural selection is the primary creative process by which evolution operates. It accounts for the entire course of evolution, from the earliest, barely differentiated organisms in the primordial ocean to the species that has had the hubris to name itself *Homo sapiens*.

Once life had begun, Earth would never be the same again. Its physical environment continued to change as elements were cycled through the crust, atmosphere and oceans; as climates changed; and as the continents were alternately built up, eroded down, and propelled slowly around the surface by the action of plate tectonics. All these processes helped determine which organisms outreproduced which others and thus the course of evolution. But the evolving life-forms also inexorably changed the physical environment of the planet.

6. **Waves and shore.** The oceans also play their part in sculpturing land; this eroded, rocky shoreline is on the west coast of County Clare, Ireland.

The earliest organisms must have gained the energy they needed by taking advantage of energy-yielding chemical reactions, possibly exploiting an abundance of organic chemicals produced in submarine hot springs by geophysical activity combined with oceanic stirring.

Opportunities to gain energy via inorganic chemical reactions are relatively rare today. But certain bacteria, members of a very ancient group of organisms known as archaebacteria, can do so, and the unusual ecosystems recently discovered around black smokers depend primarily on them.

The volcanically active young Earth's oceans, however, may have offered not only widespread suitable conditions for the formation of the primitive precursors of life but opportunities for the earliest organisms to subsist on energy from the chemical compounds continuously being formed in the ocean depths. The recent discovery that some of the black smoker bacteria could thrive at extremely high temperatures suggests moreover that life might have originated even as the planet was still cooling.

As time went on and the volcanic activity subsided somewhat, the pressure must have mounted for those first tiny organisms to find a better source of energy. At some point more than three and a half billion years ago, a micro-organism evolved that could take advantage of the most abundant and ubiquitous flow of energy on our planet — the light from the sun. The amount of energy in the form of sunlight that reaches Earth's surface daily is more than 5,000 times our present civilisation's total daily energy use; it is roughly equal to the energy yield of two million one-megaton hydrogen bombs (just one could destroy a large city). Tapping the enormous energy bonanza of the sun vastly expanded the potential of life to multiply, diversify and reshape Earth's surface.

The process by which certain organisms are able to harness the sun's energy is called photosynthesis. In this process, sunlight is absorbed by the pigment chlorophyll, and its energy is used to combine carbon dioxide and water into carbohydrates — sugars, starches and celluloses. In essence, photosynthesis converts the sun's radiant energy into the energy of the chemical bonds that hold the carbohydrates together. The familiar organisms that gain energy through photosynthesis today, of course, are green plants — from tiny mosses, roadside weeds, wheat and rice to apple, oak and giant redwood trees.

Photosynthesis developed first in microscopic single-celled organisms: bacteria. But not all organisms were able to garner energy from photosynthesis or chemical reactions; others had to acquire their energy from external sources. Some of the early bacterial organisms probably survived by attacking other living cells, consuming others' metabolic products, or by scavenging dead ones. This differentiation of lifestyles gave rise to the first rudimentary food chains.

The earliest versions of photosynthesis depended on the use of chemicals that occur rather uncommonly except in areas of undersea volcanic activity, and this must have restricted the proliferation of photosynthesising organisms. Nevertheless, these versions, still seen in certain kinds of bacteria, may have prevailed until a new type of bacteria (now called cyano bacteria; formerly blue-green algae) arose with a better method – the much more efficient form of photosynthesis used today by modern plants.

Modern photosynthesis results in a very important waste product: oxygen. Oxygen is crucial to the process by which most of today's organisms – plants, animals and micro-organisms alike – gain access to the solar energy stored in carbohydrates. That process, cellular respiration (not to be confused with breathing, which is also called 'respiration'), is a sort of carefully controlled slow burning. The carbohydrates and oxygen, both produced by photosynthesis, are combined to yield energy, carbon dioxide and water.

But rather than being given off immediately as heat and light (as would happen in the rapid burning of a fire), much of the energy in respiration is captured by a complex cellular apparatus. It can then be used to drive the life processes of the organism – maintenance, growth, reproduction and (for animals) movement and even thinking and dreaming. Some of the energy produced by respiration is directly converted to heat – which is why you continually radiate heat into your environment at about the same rate as a hundred-watt light bulb.

Respiration completes a cycle, for its outputs – energy, carbon dioxide and water – are exactly the inputs of modern photosynthesis. Green plants both photosynthesise and respire, but animals, nonphotosynthetic micro-organisms and some parasitic plants only respire and must acquire their carbohydrates directly or indirectly from photosynthesisers such as green plants. From here on, we ignore the tiny minority of organisms that gain their energy today from inorganic chemical reactions (rather than photosynthesis) and the handful of other creatures that depend upon them.

Modern photosynthesis is thought to have evolved well over three billion years ago, and it paved the way for respiration. Unlike most organisms today, the first life-forms did not use oxygen to help them extract the energy from carbohydrates because it was extremely poisonous to them. Indeed, some oxygen-containing compounds are so toxic to some of the basic chemical reactions of life that life almost certainly could not have begun in the presence of abundant oxygen – a curious irony in view of its crucial importance to nearly all life-forms today. Instead, early organisms obtained their energy from a less efficient process called fermentation, which is still employed as the primary energy mobilising process by some organisms, including yeasts which produce alcohol as an end product.

The constant release of poisonous oxygen as a by-product of photosynthesis presented a potentially serious problem for the early fermenting organisms. For hundreds of millions of years, most of the oxygen released by early photosynthesisers may have been taken up by chemical reactions with volcanic gases and with iron and other minerals dissolved in the sea. Eventually, though, such 'sinks' for oxygen were virtually used up, and the excess oxygen produced by photosynthesis slowly began to accumulate in the oceans and (after escaping from the oceans) in the atmosphere.

Sooner or later, the gradual build-up of oxygen in the environment must have produced powerful selection pressures on Earth's evolving biota – still consisting only of various bacteria – to develop ways to avoid being poisoned by it. A great advantage would have accrued to any life-forms that could tolerate its presence, and one especially good way to tolerate the presence of oxygen – indeed, turn it into a benefit – is to use it in extracting energy from carbohydrates. This was accomplished by modifying the fermentation process so the end products were carbon dioxide and water instead of carbon dioxide and alcohol (or some other organic molecule).

7. A chloroplast. In green plants and algae, photosynthesis is carried out by cellular structures called chloroplasts, which are less than a three-thousandth of an inch in diameter. One square inch of leaf may contain more than 600 million chloroplasts. The leaves of plants are green because of the green chlorophyll pigments, which are central to the photosynthetic process in the chloroplasts. The sophistication of these minuscule energy-conversion mechanisms puts that of modern petrochemical factories to shame. The cross-section of a chloroplast shown in this electron micrograph is from a leaf of a coleus plant (*Coleus blumei*). The structure in the centre that looks like a poached egg is a grain of starch, a product of the photosynthetic process, which is carried out on the thin membranes that appear as parallel lines.

The new process, cellular respiration, could be accomplished only with the addition of some complicated cellular machinery, but it yielded an enormous benefit beyond the ability to tolerate environmental oxygen. Much energy remains untapped in the organic end product of fermentation, but with respiration much more can be extracted. A fermenting organism can capture only about 2 per cent of the energy in the original sugar, while recent work shows that a

23

respirator gains over 60 per cent. Respiration is thus about thirty times as efficient as fermentation. Indeed, the 60 per cent yield from this controlled slow burning is about as high as that of the most efficient machines produced by *Homo sapiens*.

The accumulation of oxygen in the oceans thus set the stage for the evolution of new life-forms that were highly efficient at extracting energy from organic compounds. The first such organisms appeared some time before 1.8 billion years ago; by then, the concentrations of free oxygen in the oceans and the atmosphere may have reached a substantial proportion of their present levels. With these momentous changes, the ecosphere – the giant ecosystem that includes all of Earth's life-forms and the physical environment with which they interact – very gradually began taking on some of the unique characteristics it has today.

One of these characteristics is the active participation of organisms in the mobilisation of materials important to life. Some of the first primitive bacteria had developed the ability to 'fix' atmospheric nitrogen – that is, change it into chemical compounds that other organisms can use. That skill was to have far-reaching consequences for the evolution of life and to alter the physical organisation of the planet. Several kinds of bacteria still perform this service, to the vast benefit today of both natural and agricultural ecosystems, which could not persist without it. Other inconspicuous microbes break down the organic compounds and restore gaseous nitrogen to the atmosphere, thus completing the cycle of nitrogen through the interconnected living and non-living parts of the ecosphere.

The nitrogen-fixing talent appears to be very ancient, probably preceding oxygen-producing photosynthesis as well as respiration. It is especially important because nitrogen is an essential component of proteins. Critical roles are played by proteins in the structure and functions of living beings, one of which is controlling the chemical reactions of life. Long, chain-like protein molecules, called 'enzymes', fold into complex shapes inside cells and serve as biological catalysts (substances that make possible, or speed up, chemical reactions). Their three-dimensional structure helps them to regulate the rates at which other molecules join or break apart, but the enzymes themselves are not consumed by the reactions. Many of the basic enzymes commonly found even in very recent types of organisms seem to have had extremely early origins.

At some point after bacterial organisms (which are collectively known as prokaryotes) had refined photosynthesis and developed ways of coping with oxygen, a new, fundamentally different type of life-form evolved. Until then, some two billion years of evolution had produced only variations on the bacterial theme – perhaps a dozen major groups and thousands of species. The new organisms, called eukaryotes, had a much more complex and organised cell structure than do bacteria. The earliest confirmed fossil evidence for these new life-forms dates from about 1.4 billion years ago, but one rock formation about 1.9 billion years old contains fossils that may have been early eukaryotes. The first ones thus might have appeared as much as two billion years ago.

The appearance of eukaryotes was marked by advances of several kinds, starting with how these new life-forms made their livings. Whether the earliest eukaryotes were able to photosynthesise is not clear. Some of them seem to have been mobile creatures, probably rather like today's flagellates (tiny single-celled creatures propelled by whiplike 'tails' called flagella), which obtained their energy by consuming organic matter. Some of these probably ate photosynthesisers, a trend that may have led eventually to the first animals.

Evidence is accumulating that some characteristics of eukaryotes may have originated as symbiotic 'adoptions' of different kinds of bacteria, an idea promulgated by biologist Lynn Margulis. When some early flagellates consumed photosynthesising bacteria, for instance, instead of digesting them, they may have incorporated them as internal symbionts. The bacteria then continued to photosynthesise for the new, united organism and thereafter were regularly replicated when the host cells reproduced. It is now thought that chloroplasts, the photosynthesising components of cells found in all eukaryotic producers from algae to trees, may have orginated in this way. Indeed, the physical resemblance of chloroplasts found in modern plants to cyanobacteria, the bacteria with the most advanced form of photosynthesis, is striking. Similarly, mitochondria, the components that mobilise energy in nearly all eukaryotic cells, may have originated as internal symbionts.

Thus, eukaryotes may represent an evolutionary advance in that they were able to combine, use, and propagate the special characteristics and abilities of several different simpler organisms. This process of 'adopting' various symbionts may also have contributed greatly to the genetic diversity of eukaryotes and to the diversification of lifestyles.

Photosynthesing organisms are called 'producers', because the energy they acquire supports, directly or indirectly, all other life-forms. Some creatures feed directly on the photosynthesisers; some feed on others that do, and some feed on those, and so on up the food chain. All of these organisms (which today include nearly all animals) are known as 'consumers'. The third lifestyle is feeding on dead remains or excretions of other organisms; these organisms are called 'decomposers'.

25

Such decomposers perform a vital function for the perpetuation of life. In the process of digesting the complex organic chemicals characteristic of living beings, decomposers break down the chemicals into their constituent elements – essential nutrients – and return them to the environment. Producers, using energy from the sun, can then take up the nutrients and again assemble them into living tissue. Thus decomposers, also ultimately dependent on producers for their energy, are vital links in the cycles of numerous elements through the ecosphere.

An important innovation of eukaryotes was reproduction involving the regular exchange of genetic material between individuals of different sexes – that is, by sexual reproduction. Their bacterial predecessors reproduced mostly asexually, and differences in genetic information among them had primarily developed as a result of mutations – random changes in genes – possibly triggered by ultra-violet light. Genetic material is occasionally passed directly between individual bacteria, and new genetic combinations are sometimes even produced in the process, but this transfer is not associated with the primary mode of reproduction.

Most eukaryotes not only combine genetic material from two parents in a systematic way, but they usually form new combinations of their genes in the process of producing eggs and sperm, giving rise to populations that are extremely variable genetically. This variability is the raw material on which natural selection can operate – and it did. (How it does is described in the next chapter.)

The appearance of eukaryotes led to an explosive proliferation of life-forms, eventually including all the plants and animals familiar to us, many life-forms that we have never seen because they did not survive, and, of course, *Homo sapiens* itself. It is one of the surprises of modern biology, resulting from comparisons of the molecular genetic structures of different groups of organisms, that plants and animals seem to be more closely related to each other than either group is to the prokaryotic bacteria that were ancestral to all.

But all of that happened very slowly. Many hundreds of millions of years rolled by before the evolving eukaryotes developed the knack of combining cells into truly multi-cellular organisms. Bacteria and, later, algae and some primitive fungi had formed clusters or loose, chain-like associations; but multi-cellular organisation, in which clearly differentiated groups of cells were assembled into tissues and organs with specialised functions, first appears in the fossil record of about 680 million years ago.

Some scientists think this advance may have had to await the build-up of sufficient oxygen in the environment to support a multi-celled existence, because of the problem of delivering oxygen to interior cells. Because for

8. A coral reef ecosystem. The ecosystems that develop around coral reefs are among the most complex and productive on Earth. The energy that supports them is produced by photosynthesising algae that live communally with the reef-bulding coral animals. The earliest reefs seem to have been built over 450 million years ago, and the animal communities dependent on them were not very different from those found today. This one is part of the Great Barrier Reef of Australia.

aeons the accumulation depended almost entirely on photosynthesis in the seas by bacteria, it may have been exceedingly slow. By 680 million years ago, the atmospheric concentration of oxygen is thought by these scientists to have been only a fraction of that of today, although others think it was much higher by then.

Once the step to multi-cellular life was taken, though, the pace of evolution appears to have quickened. With the formation of more complex food chains, rapid differentiation among organisms may also have been stimulated by another factor: selection pressures caused by predation (see next chapter). Pressures on consumer organisms to obtain food, and on both producers and consumers to avoid being eaten, led to an extraordinary proliferation of forms and strategies to meet these needs.

The earliest animals were small and soft-bodied; the development of hard, external 'skeletons' and shells may have required a relatively high concentration of environmental oxygen. Because the hard shells prevented the simple diffusion of oxygen into cells, they were made feasible only by the development of more elaborate circulation mechanisms, which in turn could be effective only above a certain concentration of oxygen.

In the meantime, the continuing build-up of oxygen and its escape from the oceans to the atmosphere had a further profound effect on the evolution of life. Gaseous oxygen molecules consist of two oxygen atoms bound together (O_2). Such oxygen molecules may occasionally be split by ultra-violet radiation in the atmosphere to form single oxygen atoms, which sometimes then combine with oxygen molecules to form ozone (O_3). Meanwhile, ozone molecules tend to be broken down into oxygen again by other processes. Thus, when oxygen is present in the atmosphere, a small amount of ozone is constantly being both produced and broken down.

Most of the ozone is concentrated high in the stratosphere (the upper atmosphere), where relatively unfiltered ultra-violet light is able to convert oxygen to ozone. The tiny component of ozone (which today makes up only about one molecule in a hundred million in the atmosphere as a whole) is very important because it absorbs the wavelengths of ultra-violet radiation that are damaging to nearly all organisms. Water also absorbs that radiation, and until there was oxygen (and thus ozone) in the atmosphere, life had to hide under water, primarily in the oceans. Not until oxygen had built up to about a tenth of its present level, and a significant stratospheric ozone shield had formed, could survival of unsheltered life-forms on land have become possible.

The build-up of atmospheric ozone may have affected evolution in another way: as the ozone shield formed, genetic mutations must have become less frequent, increasing the importance of sexual reproduction and genetic recombination in creating new life-forms. It may also explain why some types of bacteria seem to have persisted for billions of years with little change.

Long before multi-cellular organisms had developed, the tectonic movements of the ever-changing planetary surface must have been an important influence on the course of biotic evolution. The continents, riding on top of the tectonic plates and built up over the aeons partly by their actions, have at times ground against and partially overridden one another, often creating mountain chains in the process. The velocity of the plates' movement was and is even slower than glaciers; the continents are pushed about the surface at speeds of one to a few centimetres a year. Yet, even at that leisurely pace, they have swung from one end of the planet to the other, sometimes in contact, sometimes thousands of kilometres apart.

All these continental peregrinations have inevitably had a profound effect on evolution and on the history of life. Climates have shifted in response to the continuously changing arrangements of the continents and oceans and shifts in the balance of atmospheric gases and in sea level. For instance, when continents are near the equator, their climates are warmer and more equable than when they are near the poles. Climates are also more moderate when continents are fragmented and contain in-land seas (as happens when ice caps are reduced and sea level is high) so that more of the land surface is near the oceans than when they are jammed together in supercontinents.

The proliferation of multi-celled life-forms some 700 to 500 million years ago seems to have been favoured by the conditions existing then – an abundance of warm, shallow seas and a generally benign climate. By 500 million years ago, less than 200 million years after the debut of multi-celled organisms, representatives of nearly all the major groups of animals existing today – including various marine worms, molluscs, arthropods and primitive fish (the first vertebrates) – had appeared. Eukaryotic producers included a variety of seaweeds – algae, diatoms etc. – of the same general types seen today.

Fossils have also been found of a number of strange and otherwise completely unknown marine animals, which apparently represent evolutionary experiments that failed. When and why they disappeared is not known. Conceivably, a widespread change in climate resulting from continental motions, or an outbreak of volcanic activity, could have made the environment inhospitable for those life-forms. Or perhaps the ancestors of today's organisms were simply more successful in the competition for resources.

By 450 million years ago, the first coral reefs had developed, and marine communities had assumed more or less their present form.

29

Most of the principal groups of flora and fauna in the oceans had appeared, and their ecological relationships had been pretty well established. Changes in marine ecosystems since then have been more in detail than in overall structure, as the various groups have evolved further and produced new species. The chief scene of evolutionary action had shifted to the land.

Photosynthesising organisms were necessarily the first to colonise land surfaces permanently, since animals and decomposers are dependent on them for their energy and nutrients. But exactly when and in what form the first terrestrial producers appeared is not known; no fossils have been found to shed light on their beginnings. Certain green algae were the first eukaryotic colonists (possibly preceded by bacteria) and they evolved into the first primitive terrestrial plants. Indirect geological evidence suggests that some form of photosynthesising life may have existed on land, in wetlands and along streams, as much as 700 million to a billion years ago. But the earliest solid evidence of primitive vascular plants (those with differentiated stems, branches, and later, roots) appeared just over 400 million years ago.

9. **Insects and green plants.** Among the first multi-celled organisms to live on land were primitive green plants and insects, which for several hundred million years have been engaged in a 'co-evolutionary race' with each other; the plants to defend themselves against the attacks of herbivorous insects and the insects to overcome the defences of the plants. This modern grasshopper represents one of the earlier insect groups to evolve.

Once plants had gained a foothold on land, animals were not far behind. The first to venture ashore seem to have been snails, round worms and arthropods – insects, spiders and mites. Fossils of some of these creatures are even older than the earliest ones of vascular plants (though the presence of the animals clearly implies the prior existence of terrestrial producers on which they could have fed).

After plants and animals invaded the land, they proliferated and spread, partly under the influence of barriers erected and removed by continental movements and associated climatic changes. Climatic factors doubtless assumed an even greater evolutionary importance for life on land than they had for marine communities. Differences in temperature due to latitudes and seasons are particularly pronounced in continental interiors, because land areas respond more quickly than oceans to heating and cooling. Moreover, the patterns in which rain and snow fall on land govern the availability and abundance of water in any given place and are therefore critically important in shaping biotic communities.

The evolutionary paths of terrestrial plants and insects seem to have been especially closely interconnected, each group exerting selective pressures on the other – the insects feeding on the plants, and the plants devising ways to avoid being demolished – a kind of interaction known as coevolution. The result seems to have been a rapid diversification of both groups.

At first the plants were fairly modest in size, but by about 350 million years ago, some had attained the size of small trees. Ferns and their relatives dominated the lush, moist forests of the time, but the first seed plants, relatives of contemporary conifers, soon appeared. The proliferation of plant life on land probably led to an enormous boost in global photosynthetic activity, which may have further increased the concentration of oxygen in the atmosphere.

By then, another type of animal had entered the scene. The first terrestrial vertebrates, probably in the form of a fish that could survive for a time out of water, crawled out of the sea around 350 million years ago. Their descendants were amphibians, which could live on land, but still needed plenty of moisture to survive and an aquatic

10. **Dinosaurs.** The dominant form of animal life on land between 200 and 65 million years ago was reptiles, primarily the dinosaurs, some of which would have dwarfed today's elephants. Most of the early dinosaurs were herbivorous; later, carnivorous forms evolved that preyed on their vegetarian cousins and doubtless on another group of animals, the mammals. This artist's conception is of Iguanadon dinosaurs, which were herbivores and lived in the reed swamps of the Sussex Weald about 130 million years ago.

environment in which to reproduce. The next step was development of an egg with a shell that could withstand a dry environment – an innovation of the reptiles, the first of which seem to have arisen about 300 million years ago.

The first amphibians and reptiles appear to have been mainly insect-eaters, thus adding another link to the diversifying terrestrial food chains and food webs. (Most food chains are interlinked in complex ways; most plant-eaters eat more than one kind of plant, animal-eaters eat more than one kind of animal; each usually has more than one predator.) The early terrestrial vertebrates were relatively small; the heyday of giant dinosaurs was yet to come.

The age of dinosaurs began around 225 million years ago and lasted for 150 million years. Many dinosaurs were herbivorous (plant-eaters), unlike their insectivorous ancestors, and some preyed on other reptiles and amphibians. The prevailing climate seems to have been warm and equable, favouring the continued evolution and increasing dominance of seed plants and, eventually, the appearance of the first flowering plants. Also during this period, the first mammals and birds arose; but both groups remained relatively inconspicuous while the dinosaurs prevailed.

The long history of life has mostly been a tale of a steadily increasing variety of organisms; new species have appeared, and, on a time scale of a million years or so, disappeared to be replaced by others. Nothing happened very fast, apart from a few dramatic occasions in which a substantial portion of Earth's complement of life-forms went extinct quite rapidly. For instance, about 65 million years ago, the last remaining dinosaurs, then the dominant land animals, vanished forever from the fossil record, along with some important groups of plants and many marine organisms.

The exact cause of that episode of widespread extinction, and even the speed with which it took place, are difficult to determine from the fossil record and have been the subject of much controversy. It could have occurred in slow motion from a human perspective, since it might have been caused by a large-scale change in climate stretched over tens or hundreds of thousands of years, resulting from the slow tectonic motions of the continents. Or it might have occurred more rapidly as a consequence of increased volcanic activity. But some scientists have recently pointed to evidence of a great catastrophe caused by a collision of Earth with an asteroid or some other celestial body.

Such a collision could have lofted huge quantities of dust into the atmosphere for a substantial period, which would have blocked out the sun's rays, appreciably cooled the planet's surface, and deprived organisms of their basic energy source by shutting down photosynthesis. An event of that magnitude clearly could have decimated Earth's flora and fauna by causing the disappearance of myriad life-forms that were sensitive to the cold or lacked sufficient energy reserves to survive a long period without sunlight or food.

Further in the past, other massive extinction episodes appear to have occurred, though even less is known about their causes. One major event was about 230 million years ago – about the time that all the continents merged into a single giant supercontinent called Pangaea. The massive convergence of all the continents and the disappearance of the seas that had previously separated them surely caused marked changes in climates and habitats, and possibly increased volcanic activity. Most likely, climate changes resulting from Pangaea's formation would have come on gradually, but volcanic activity might have caused relatively sudden, severe changes.

11. **The San Andreas fault.** The geological fault seen clearly in this photograph marks a boundary where two of the tectonic plates that comprise the surface crust of Earth are grinding against each other, producing frequent earthquakes. The San Andreas fault runs south-east in California from the Pacific shore near San Francisco into northern Mexico. This fault produced the disastrous San Francisco earthquake of 1906.

The evolving biosphere ultimately recovered from the extinction events, generating new communities and new states of equilibrium. The loss of the dinosaurs thus opened the way for a proliferation of mammals, which had formerly been overshadowed by their giant reptilian cousins. The era of mammalian dominance was accompanied by the flourishing of today's most important group of plants – the

33

flowering plants. Whatever the cause of the extinctions, the eventual result was an appreciably different collection of life-forms on the ever-changing Earth.

12. **A glacier in Paradise Bay, Antarctica.** Often in Earth's history, glaciation has played a major role in shaping land surfaces and influencing climates. The isolated island continent of Antarctica was not always near the South Pole; some 250 to 200 million years ago it was part of Pangaea, a giant supercontinent formed when all the continents clustered together for a hundred million years or so. When Pangaea first formed, about 275 million years ago, Antarctica is thought to have been south polar and covered with a giant ice sheet. Pangaea then drifted gradually northward. Fossils indicate that, as Pangaea began to break up, around 200 million years ago, the climate in Antarctica was mild, even equatorial.

Tens of millions of years later, a drastically different climatic age, characterised by gigantic advancing and retreating glaciers covering vast continental areas in the north temperate regions and Antarctica in the south, stimulated further changes in terrestrial flora and fauna. Climates became much more disparate from place to place, and more changeable over time, than they seem to have been earlier. The variety of habitats shaped by these climatic disparities may have given rise to a dramatically increased diversity of life-forms in the past 40 million years or so.

The stage was set for the entrance of a new species of mammal, one that walked on its hind legs, communicated verbally, and rose to a new kind of dominance. But before examining the implications of the new arrival's impact on the living planet that produced it, let us look more closely at the character and workings of the world it inherited.

CHAPTER TWO

The Legacy

In the four billion or so years since the first primitive cells formed in the early oceans, Earth has been gradually but profoundly transformed as life evolved. The changes seem to have been so slow at first that space travellers visiting the planet at long intervals during the first three billion years might have noticed few differences, except perhaps in the sizes and shapes of the continents and oceans. But in the last several hundred million years, especially since the first organisms ventured out of the sea, the pace of change has accelerated with the diversification of life-forms.

Indeed, one might speculate that, once made habitable by the ozone shield, the land has proved considerably more hospitable to life than its wet and salty birthplace. Even though barely a quarter of the planet's surface is land, about three-fifths of the world's biological productivity today (measured as the carbohydrates produced annually by photosynthesis) is generated there, and by far the greatest diversity of organisms is to be found in terrestrial ecosystems.

It was this extraordinary, life-bearing, life-sustaining planet that ultimately produced and shaped humankind. This chapter explores the dimensions of that unique inheritance, both its physical and its biotic aspects. The former include the deposits of minerals that made civilisations possible. Among the latter are the workings and structure of the ecosphere – the interacting living and non-living components of the planet – as well as the process, evolution, that has brought life to its present level of richness and diversity. As will become apparent, the living and the non-living components are not always easy to distinguish.

Far from being a static lump of clay orbiting its undistinguished star, Earth has always been a dynamic entity, even apart from the evolution of life and the profound changes that it has wrought. Since the first cooling, the crustal surface has shifted, slipped, fractured, boiled up material and thrust it underneath again, all the while sliding over the plastic mantle layer, which in turn overlies the planet's molten core. Crustal land masses gradually emerged as the early oceans filled up.

Ever since, the continents have been constantly pushed around the surface and built up by tectonic action, while being inexorably worn down by processes of weathering and erosion.

All this geophysical activity resulted in continuous, if slow, redistribution of the materials of which the planet is made up. In some cases, it produced concentrations of certain minerals, for instance gold, silver, copper, iron and other metals, a circumstance that eventually proved of enormous value to human beings. Often the richest concentrations of materials have turned up in mountainous areas that mark the sites of ancient 'seams' where the continents were stitched together by tectonic collisions. One such region is the so-called 'overthrust belt' of the northern Rocky Mountains in the United States, where two land masses were joined hundreds of millions of years ago.

Water, too, is continually being redistributed. The overwhelming bulk of Earth's water, of course, is in the oceans. But, virtually from the beginning, water has cycled through the atmosphere, evaporated from wet surfaces by the sun's energy, moved with the atmosphere's circulation from place to place, and returned to the surface in precipitation (rain, snow, hail, sleet). Much of the pure water that falls as rain on land surfaces runs off into rivers and lakes, and eventually finds its way back to the sea. But a substantial proportion of it – especially since the development of terrestrial plant life, with its capacity to generate soil and hold it in place – seeps deep into the ground. There the water is stored in natural underground reservoirs, spaces in and between rock formations, which are called aquifers.

Vast amounts of fresh water have filled these aquifers underlying the continents and islands around the world – a quantity many times greater than that found on the surface in rivers and lakes at any given time. The underground water also circulates, but at an extremely slow rate compared to that on the surface and in the atmosphere. Underground streams and springs return water to the surface and slowly carry it back to the oceans, while fresh water constantly percolates downward from the surface to replenish the aquifers. Part of this enormous ancient store of underground water is the source that people tap in wells; today it is an increasingly important part of our inheritance.

Over the aeons, as Earth and its biosphere (the living portion of the ecosphere) evolved, organisms gradually became more and more involved in the distribution and cycling within the ecosphere of all the elements critical to life (known in this context as nutrients). This involvement of life-forms with some nutrients – such as nitrogen, described in Chapter 1 – began even before photosynthesising organisms started changing the distribution and forms of oxygen, being responsible for the existence of free oxygen in the atmosphere and the oceans.

13. Plants and weather. The cycling of water from oceans to atmosphere (as water vapour) to land surfaces (as rain and snow) and back to the oceans (as rain or in rivers) is driven by energy from the sun. But rainfall patterns on land are also profoundly influenced by the vegetation there, in part because most plants continuously pump water from the soil into the atmosphere.

Carbon, like nitrogen, is a nutrient that circulates in a great global cycle. Photosynthesising organisms 'fix' the carbon from carbon dioxide present in the oceans or the atmosphere, incorporating it into carbohydrates. Some of that carbon is returned to the environment as carbon dioxide by the respiration of the producer organisms, and some is acquired by herbivores (plant-eating animals) when the plants are consumed. The herbivores, in turn, return some to the atmosphere through their respiration and yield some to carnivores (animals that eat animals), which also return carbon to the atmospheric pool through their respiration.

At every step in food chains, including the producers and each of the consumer links, the constituent elements (of which carbon is one) of dead organic matter are restored to the environment by the decomposers. Organisms die; they excrete; they shed leaves, bark, eggshells, skin or hair. And myriads of decomposer animals, fungi and micro-organisms make their livings by extracting the remaining energy from the debris. Through their respiratory processes, they too contribute carbon dioxide to the atmosphere.

This great carbon dioxide → photosynthesis → organic molecule → respiration → carbon dioxide circularity of organic carbon is only one of the global cycles of materials in which Earth's life-forms participate.

37

14. Photosynthesis. Green leaves, like these in a tropical rain forest in Brazil, capture energy from sunlight through the process of photosynthesis and make it available to all the other organisms in an ecosystem.

(Carbon also moves in a cycle outside the biosphere from weathering of rock to ocean to atmosphere as carbon dioxide.) Oxygen and hydrogen, which along with carbon are the other constituents of carbohydrates, cycle through the biosphere virtually in step with the carbon. A number of other materials needed by organisms for their life processes, notably the important nutrients nitrogen and sulphur (key components of proteins) and phosphorus (an essential element in compounds that transport energy) also travel through such cycling pathways.

Usually, the role of organisms has been to help make essential nutrients more available to themselves and each other (as with nitrogen-fixing bacteria). Sometimes, though, the effect has been to lock materials away from the biosphere. Photosynthesising organisms over the ages, for instance, have produced slightly more organic material than was consumed or decomposed by other organisms. Instead, a vast amount of undecomposed organic material has accumulated under the mud on the floors of great swamps and shallow seas, out of reach of oxygen.

As a result, a large accumulation of carbon was never oxidised to carbon dioxide, and the oxygen produced in photosynthesis was therefore able to build up even faster in the oceans and atmosphere. Meanwhile, the carbon-rich buried material was transformed over tens to hundreds of millions of years by geophysical processes into peat, coal, natural gas and petroleum. Thus, some of the energy that arrived from the sun in the distant past was stored, eventually to provide part of the rich inheritance for our civilisation. (The origin of some petroleum and natural gas is currently a matter of controversy. A few earth scientists believe that vast, untapped reservoirs of these fuels of non-biotic origin exist deep in Earth's crust. Most geologists disagree, but explorations are now under way.)

Unlike nutrients, energy, once used, cannot be recycled. Each time organisms use energy, part of it is transformed into a less useful form. This loss of usefulness of energy is inevitable in real-world processes. This universal property (to which no exception has ever been observed) is described by the famous 'second law of thermodynamics'. It is this property of energy that makes perpetual motion machines impossible and guarantees that less of the sun's energy will be available to plant-eaters than is available to plants, still less of that energy will be available to the animals that eat the herbivores, and so on up the food chain.

The establishment of photosynthesis as the primary energy source of life, the rise to dominance of respiration as the mechanism for mobilisation of that energy, and the accompanying establishment of global nutrient cycles with biotic components, completed the outline of the structure of the biosphere as we now find it. But many other processes have been involved in filling in that outline. Prominent among them has been the one responsible for generating the great diversity of kinds – species – of organisms existing today: evolution.

Evolution operates mainly through natural selection, a process by which populations can change from generation to generation. Individuals of some genetic types (genotypes) reproduce themselves more successfully than others. The less successful genotypes are, by definition, 'less fit' in the environment of the moment. As an environment changes, so does the fitness of genotypes living in it, and, as a result, so do the genetic characteristics of the population.

We can see it happening today. For instance, an insect called the peppered moth lived in Great Britain in the early nineteenth century. As its name implies, this moth was white with dark spots and blotches. When it rested on the lichen-covered bark of trees, it was almost invisible to birds (which like to eat moths). Occasionally, a melanic (dark, almost black) moth would be produced through a random change, a mutation, in the genes of one of its parents. The dark

39

mutants stood out like brightly coloured road signs against the pale lichens, and hungry birds quickly gobbled them up. The melanic genotypes were less fit than the peppered ones.

Then along came industrialisation, and in the British Midlands lichens began to die from air pollution, which also coated tree trunks with grime. Soon the melanic moths were the camouflaged ones and the normal peppered moths stood out. The fitnesses were reversed, and the melanic genotypes subsequently out-reproduced the peppered ones. In this case, differential survival caused differential reproduction. The frequency of peppered individuals in the polluted Midlands declined until they persisted only as rare mutants, just as the melanics, which now make up the bulk of the population, had earlier. Interestingly, though, the area containing a high proportion of melanics has shrunk in the last twenty years as control of smoke emissions has again changed the moth's environment.

This example of natural selection shows how evolution, in response to environmental change, can transform a population. But, to return to our previous question, how was evolution responsible for the great diversity of species found today? Basically, it is because the physical environment has always varied from place to place and from time to time. If the surface of Earth had always been absolutely uniform, the five to thirty million different kinds of plants, animals and micro-organisms now existing could not have evolved. But there have always been shallow and deep seas, turbulent and still waters, sandy and rocky shores, hot and cold continental areas, wet and dry slopes of mountains, and so on. Natural selection has thus operated differently in different places; genotypes that were most fit in one spot were less fit in others – just as the fitness of the peppered moth genotypes varied with their distance from sources of pollution.

One kind of organism that lived in two different physical habitats would therefore be subjected to two different sets of selection pressures, and the populations in the two areas would gradually diverge in their characteristics. If they remained separated under different conditions for long enough, they would diverge so far that, were their populations ever rejoined, they would be unable to interbreed. Thus two different species would live where only one had been before; the process of 'speciation' would have occurred.

Although this account of speciation is very much simplified, it captures the essence of the process that biologists believe to be responsible for the extraordinary diversification of life on Earth. Geographical differences in physical environments result in speciation, which then produces different communities of plants and animals in physically different places. Because the plants and animals in a community are part of each others' environment, selection pressures

are even more diverse from one location to another, further
encouraging speciation.

It would seem that speciation, as just described, would quickly fill
Earth with species to overflowing. But, as Charles Darwin first
recognised, the process of extinction tends to counterbalance the effects
of speciation. Some species, through better ability to compete for
resources, a capacity to evolve rapidly when confronted by
environmental change, or sheer luck, persist for long periods; others
die out. Biologists estimate that roughly 98 per cent of all the species
that have ever lived are now extinct.

Extinctions have not occurred steadily over the long history of life.
As mentioned in Chapter 1, extinctions seem to have been
concentrated (as far as can be told from the sketchy and sporadic fossil
remains) in a handful of catastrophic episodes. Not only is there
controversy about the causes and speed of past extinction episodes,
there is also debate over whether the pace of speciation has been
relatively steady or stuttering. Have new species always been forming
(as they appear to be now)? Or has much of speciation, like much of
extinction, been concentrated into a few relatively brief periods such as
the remarkable burst of diversification that seems to have followed the
multi-cellular breakthrough? Nobody can say for sure, again because of
difficulties in interpreting the fossil record.

What is sure is that speciation, on average, has kept ahead of
extinction and produced the luxuriant panoply of life with which we
are familiar. It has produced extraordinarily rich communities of
organisms, wherein each species is intricately adapted to its physical
environment and to the other kinds of organisms with which it shares
its habitat.

Sometimes the adaptations of organisms to each other have
developed through the kind of reciprocal evolution called coevolution.
Intimate associations of several kinds can exist between pairs of
species, such as a herbivore and its food-plant, a predator and its prey,
or mutualistic relationships such as lichens (a mutually dependent
association of an alga and a fungus).

In coevolutionary associations, each species exerts selective pressures
on the other. The carnivore becomes increasingly efficient at hunting
its prey; in response, the prey tends to become ever more adept at
eluding the predator.

Similarly, the herbivore tries to improve its efficiency in finding and
eating its food; the plant in turn must defend or protect itself from
being devoured. Plants cannot run away, but they can – and do – hide,
disguise themselves, bloom or set seed when the herbivores are not
around, or use unpleasant or poisonous chemicals as defences. Indeed,
the defensive chemicals produced by plants are among the most useful

41

15. **Industrial melanism.** The mottled pattern of the peppered moth (*Biston betularia*) camouflaged it effectively against the lichen-covered bark of trees until the acceleration of coal burning in the early industrial revolution began to blacken tree trunks with soot and kill the lichens. In areas with sooty trees, the nearly black melanic form of the moth (right) held an evolutionary advantage compared with the peppered form and replaced it. Today, with less pollution, tree trunks are returning to their original colour, and more of the peppered forms of the moth are surviving.

16. **Sexual reproduction.** In sexual reproduction, genetic material from the parents is recombined when sperm or eggs are formed in an offspring. This regular reshuffling of genes gives rise to genetic diversity within a population. The copulating pair here are butterflies (*Plebeius acmon*) in Colorado. Today, sexual reproduction is much more common than asexual reproduction, but the evolutionary advantage of sexual reproduction remains a subject of intense technical debate.

17. **Diversity of plant life.** Flowering plants, now the dominant form of terrestrial plant life, have diversified over the past 100 million years or so to some 250,000 species. These lovely wild-flowers, including bluebells and greater stitchwort, are growing on a bank in Sussex.

18. **Predators and prey.** This lion is bringing down an African buffalo. These predators and their prey have coevolved for millions of years. A lion's attack fails at least as often as it succeeds. Like many predators on hooved animals, it most often catches very young, very old, ill or injured individuals.

items to turn up in the human legacy, including spices and flavourings, stimulants, drugs, antibiotics, insecticides, rubber, oils and so on.

Coevolutionary relationships have doubtless played a large role in shaping most biotic communities. Jointly, these communities and the non-living environments with which they interact make up the ecosystems of the planet. And, in turn, ecosystems have greatly modified their physical environments in numerous ways, including the generation and maintenance of the vast nutrient cycles that permit the survival of the systems themselves.

The very mixture of gases in the atmosphere, as we have seen, is largely the creation of, and is still partly controlled by, the activities of living organisms. The balance of the two major atmospheric components, nitrogen and oxygen, is of particular significance. Everyone realises that an appreciable drop in oxygen content would be disastrous, but few people are aware that even a small *rise* in oxygen content would also be catastrophic. If the concentration of oxygen rose from 21 to 25 per cent, even the wet tropical rainforests might burn vigorously, wildfires might destroy all vegetation on land, and most life-forms might be able to persist only under water.

But oxygen is only the most obvious of the gases in the atmosphere with biotic connections. Also present, in minute amounts, are a number of 'trace' gases, such as methane (natural gas), hydrogen, and nitrous oxides, which are also intimately related to biological processes. And the concentration of carbon dioxide, currently about 350 parts per million in the atmosphere (or .035 per cent), is largely controlled by organisms, especially green plants, as part of the carbon cycle.

43

19. **Herbivores and plants.** This giraffe browses on acacia trees in the African savanna. Other large herbivores usually cannot reach leaves as high on the tree as can giraffes. Many herbivores are also deterred by the acacia's thorns, an important defence mechanism of the plant.

20. **Herbivores.** The largest herbivores today are the elephants, one species of which now survives only in Africa and another in a few places in southern Asia, although several species existed 30,000 years ago and occupied most of the continents. This cow African elephant and her calf are in Lake Manyara National Park in Tanzania, a protected area.

Carbon dioxide is a prime component of the so-called 'greenhouse effect' which warms the air in the lower atmosphere and maintains surface temperatures at levels hospitable to life. Carbon dioxide is relatively transparent to visible light, allowing sunlight to reach Earth's surface nearly unhindered, but it tends to absorb radiation in the infra-red range — the form in which part of the sunlight (after absorption by the surface) is radiated back towards space. Much of the infra-red radiation is thus trapped and held by the carbon dioxide in the atmosphere.

Water vapour, like carbon dioxide and some of the trace gases, also affects the atmosphere's heat balance. As vapour, water is partially transparent to the incoming sunlight, but when it condenses as clouds, it blocks a portion of the sunlight from reaching the surface, absorbing some and reflecting much of it back to space. Both clouds and water vapour also absorb and re-radiate much of the outgoing infra-red radiation. Water in the atmosphere thus contributes to the greenhouse effect. Without the insulating capacity of these atmospheric components, the average temperature of Earth's surface would be below freezing.

Ecosystems play other roles in regulating and ameliorating climate (average weather conditions) beyond their contributions to the gaseous make-up of the atmosphere. The sun is the power plant that drives Earth's climate. Some areas of the planet's surface absorb more sunlight than others and are warmed as a result. The air above those relatively warm places is also heated, and it rises, cools, and then descends over relatively cool areas. Meanwhile, the rising warm air must be replaced, and cooler air is drawn in from nearby areas, creating winds. If the warm air contains water vapour, rising and cooling leads to condensation (forming clouds) and often to precipitation.

This general process underlies the major atmospheric circulation patterns and climatic phenomena of the globe, as well as driving local weather on a smaller scale. For instance, warm, moist air generally rises over the hot, wet equatorial areas, and, as it cools and expands, the moisture in it condenses and falls as precipitation. Cooler, drier air flows along Earth's surface towards the equator to pick up moisture and be warmed in turn. High in the atmosphere, dry, progressively cooler air flows toward the poles, and at about 30° north and south latitude it descends, warming by compression, to replace the air moving towards the equator along the surface. This is the basis of the general circulation pattern between the equator and mid-latitudes. The descending dry air in sub-tropical latitudes (around 30° north and south) creates the characteristic bands of deserts found there.

The tidy circulation pattern just described is distorted by Earth's

21. **A savanna ecosystem.** Savannas are grassland systems dotted with trees that are found in semi-arid tropical and subtropical regions. In Africa, they often support a large, diverse fauna of herbivores, such as the migrating wildebeeste seen here in the Serengeti Plain of East Africa. Besides an enormous herd of wildebeeste, this broad plain is host to zebras, cape buffalo, giraffes, warthogs and several species of antelope. These animals 'share' the plant resources by specialising on different species or plant parts in ways that ordinarily do not damage the productivity of the savanna.

22. A desert ecosystem. Deserts are relatively arid systems, typically found in a belt around 30° north and south of the equator. Although water is scarce, the Sonoran Desert of Arizona nevertheless supports a wealth of plant and animal life. Among the plants seen here are brittle bush, owl's clover and cholla, saguaro and organ pipe cacti.

23. An extreme desert. Even exceedingly dry areas, such as this part of the Namib desert, 200 miles south of Walvis Bay, Namibia, can support a surprising diversity of plant and animals species, which are adapted to the extreme conditions. In some parts of the Namib, rain may fall only once or twice in a decade.

rotation, which causes parcels of air moving towards the equator in the tropics and sub-tropics to be deflected westward. Wind patterns are also distorted near the surface by friction, which slows the motion of the air and changes wind direction. Circulation is also affected by the arrangements of the oceans (whose currents transfer heat) and continents (which heat up and cool down more rapidly than oceans). Wind patterns are disturbed too by mountain ranges (which block air flow and cause rising air and turbulence as winds blow over them). Finally, patterns of airflow are strongly influenced by cloud cover as it reflects away incoming solar energy and traps energy being radiated outward by Earth's warmed surface.

The detailed workings of the global circulation patterns just outlined need not concern us here. The important point is that, on both global and local levels, ecosystems help control climate and weather by influencing patterns of heating and air flow. A key factor in the climate of any area is the reflectivity (technically called the 'albedo') of the surface – that is, the proportion of incoming solar energy that is reflected rather than absorbed by the surface. Different surfaces have very different albedos: snow reflects 50–90 per cent of the sunlight that hits it, clouds 25–90 per cent, sand 20–30 per cent, forests 5–25 per cent, and so on.

Green plants can exert a powerful influence on the albedo, both directly, by their own reflective characteristics, and indirectly, by influencing the atmosphere's moisture content. For example, the heavy cloud cover of the Amazon basin is partly generated by its forests. Moisture originating over the Atlantic ocean falls as rain on the forest, where it is quickly taken up by trees and other plants and pumped back into the atmosphere in a process called 'transpiration', by which plants take up water through their roots and release it as vapour into the atmosphere, largely through pores in their leaves, as a by-product of respiration. There it forms clouds and again descends as rain, only to be injected once more into the atmosphere.

The cycle is repeated over and over; an average molecule of water originating in the Atlantic passes through the vegetation and atmosphere about nine times as it crosses the Amazon basin. Were it not for the transpiring forest, there would probably be much less cloud over the basin. Without the clouds, the albedo would be markedly reduced, and the region's climate would be noticeably drier and hotter.

As this example suggests, ecosystems (especially forests) are extremely important in regulating the cycle and flow of Earth's fresh water. And soil ecosystems can absorb and, to a degree, purify water. Rainfall is soaked up by both the plant cover and the soil, then gradually released over days or weeks through springs and streams. Some of the water sinks deep into the ground to recharge aquifers.

Soil itself is continuously produced and enriched by natural ecosystems. Earth's continents are continually being worn away by the scouring action of winds, flowing water and glaciers, as well as by temperature changes, ice forming in cracks, the invasion of roots, and other processes that gradually fragment and crumble rock. Before life had colonised the land, soils were only pulverised rock, unprotected against the eroding forces of wind and water.

Now, however, soils are complex ecosystems in themselves, in which the fragments of rock are mixed with the remains and waste products of organisms to form the non-living components. The enormously diverse living components of soil are essential to its fertility. Few people realise that healthy soils contain a wonderland of tiny organisms making a living for themselves and, in the process, making life possible for all the other organisms on land, including people. Exactly how these extraordinarily complex systems evolved is not known, but the present-day result, built up over millions of years by generations of green plants, tiny animals and soil microbes, is also the indispensable foundation of agriculture.

Every secondary school should give its students the following instructive exercise: place a handful of soil from a forest or a field (not a recently pesticided farm field) in a funnel with a screen in its bottom; allow a light bulb to burn at the top of the funnel, and place the funnel's spout in a bottle containing alcohol. As the soil slowly dries, some of the myriad tiny organisms in it — small worms, mites, insects, etc. — will migrate downward and drop into the alcohol. After a few days, using a low-power microscope, students can examine the contents of the alcohol bottle. Never again will they view soil as simply lifeless dirt.

A gram of rich agricultural soil has been shown to contain more than 80,000 single-celled protists (algae, protozoa, etc.), 400,000 fungi and 2.5 billion bacteria. Beneath the surface of a square metre of Danish pasture soil were found some 45,000 minute relatives of earthworms, 48,000 mites and insects and 10 million round worms.

These inconspicuous soil micro-organisms play vital roles in operating the nutrient cycles on which all terrestrial organisms (including ourselves) depend. One important example is the presence in soil of the ancient types of bacteria that can 'fix' nitrogen — change the atmospheric gas to a form that is usable as a nutrient by plants and other organisms. Some of these inconspicuous microbes often live in intimate association with the roots of plants in the legume family (peas and beans). In return for energy-rich products of the plants' photosynthesis, the bacteria enrich the soil with nitrogen.

Similarly, the presence of certain soil fungi enables many plants to take up the nutrients they require. For instance, while the 'dominant'

49

24. **A mist in the rain forest.** An enormous amount of water is cycled continuously between a moist tropical forest and the atmosphere, as can been seen by the early morning mist shown here in the Santiago Valley of Equador. Tropical rain forests like this one are among Earth's most complex ecosystems, harbouring the richest diversity of life-forms.

25. **Decomposers.** These Rüppell's vultures feasting on the corpse of a zebra, which was killed by a large predator, are among the largest and most conspicuous decomposers. Most decomposers are bacteria, fungi, small insects, mites or worms. Along with their tiny colleagues, the vultures perform a vital service in ecosystems by disposing of carcasses and, in the process, recycling essential nutrients.

26. **An intertidal area.** The areas between high and low tides along shorelines are places where the environment is constantly changing and is frequently disrupted by storms. They are also areas of great biotic diversity and were probably major scenes of action when early multi-celled organisms were evolving. The intertidal zone here is in the Scilly Isles.

51

plants in a forest appear to be trees, the true dominants may be so-called 'mycorrhizal fungi' living in and around the trees' roots. In exchange for carbohydrates, the fungi transport nutrients from the soil to the roots.

The dependence of the trees on the fungi can be demonstrated by experiment. If the seeds of white pines are germinated and the sprouts grown in nutrient solution for a few months, and if the seedlings are then transplanted to certain prairie soils that lack mycorrhizal fungi, the seedlings will die of malnutrition. But if they are first grown in forest soil and then transplanted to prairie soil, they grow splendidly. The reason is that they acquire from the forest soil and carry with them the fungi necessary for their survival.

Soils are constantly being diminished as wind and water wear them away, but the rock particles, nutrients and important structural components are also constantly replenished by the slow fragmentation of the parent rock, the reproduction of soil organisms and complex chemical interactions among the organisms, their organic remains and inorganic components from the rock. Replenishment by these processes is continuous, though it generally proceeds at an exceedingly slow rate – a rate best measured in centimetres per millennium. That gradual renewal rate is sufficient for natural ecosystems because living parts of the soil system, especially the roots of plants, serve to anchor the non-living parts and thus maintain the necessary physical substrata for the entire soil complex.

27. **Seal.** In Antarctica's harsh climate, virtually no plant life exists on the land, and most photosynthetic productivity occurs in the sea. Most of the animals that reproduce on land, such as this Weddell seal, must find their food in the sea.

Many of the tiny organisms in soil, as well as much larger organisms such as lions on the Serengeti plain and condors soaring over the coastal mountains of California, perform another important function in ecosystems – waste disposal. These are the decomposer organisms, ones that live entirely or in part on excretions or on the corpses of organisms they did not kill (lions often steal carrion from smaller predators).

The role of decomposers in ecosystems is to break down organic wastes into the basic elements – such as nitrogen, carbon, hydrogen, oxygen, sulphur, iron etc. – of which they were originally made. These nutrient elements then become available for use again by other organisms, primarily plants, which can take them up from the soil or the atmosphere. From the plants, the nutrients are taken by herbivores, then passed through the rest of the food chain.

Breakdown of organic wastes is similarly carried out in aquatic systems by bacteria and other organisms. This capacity for waste disposal keeps rivers and lakes to a large extent self-cleaning and tolerant of a moderate degree of pollution. Thus decomposers, by the processes of decay and decomposition, complete Earth's cycles of nutrients, ensuring that those essential elements of life are continuously supplied for its perpetuation.

The functioning of natural ecosystems – cycling of nutrients, metering of water, etc. – is generally quite consistent, even though populations of organisms often fluctuate enormously in size, and extinctions of populations may occur frequently. This stability is thought by some biologists to be related to the complexity of a system and the diversity of its resident species, although some rather simple systems are quite stable and some complex ones are highly vulnerable to many types of disturbance.

Probably one reason for the functional stability is redundancy. More than one kind of organism may perform each role in the system, thus ensuring that functions will be carried on even if one of the organisms disappears. Feedback systems also operate; an outbreak of one population may lead to a population surge of one or more of its predators, which soon restore control of the first population.

Moist tropical forests are exceedingly complex ecosystems. There, various forms of coadaptation, cooperation and mutualism between species, perhaps resulting from a long, comparatively uninterrupted period of coevolution, have reached a level of fine-tuning not seen elsewhere. At the same time, competition for nutrients and other resources (even sunlight) in tropical systems is intense. This combination of the intricacy of the system and the tenacity of individual species may account both for the apparently high resilience of these systems (that is, their capacity to recover from minor

28. **Penguins.** Like the Weddell seal, Antarctic penguins feed in the sea. Here an adult Adelie penguin feeds its chicks by regurgitating krill (small shrimp-like crustaceans) brought back from a hunting expedition. Note the adult's spined tongue, which aids it in gripping the krill. The parents take turns hunting and caring for their young. Penguins could probably survive in the Arctic. The now-extinct Great Auk (which was the first bird to be called 'penguin') lived in the North Atlantic and filled an ecological role very much like that of Antarctic penguins. But the latter have never successfully crossed the barrier of the warm tropical oceans. The furthest north they have gone is the Galapagos archipelago, where the cool waters of the Humboldt current reach the equator.

disturbances) and their fragility (their inability to restore themselves after severe disruption).

Ecosystems themselves vary greatly in character as well as in the details of their composition – the particular community of species in each one – reflecting great differences in climate, solar radiation, availability of water and nutrients, and so forth. Consider the extremes. Very arid deserts, such as the Sahara, and arctic deserts, where a cold climate combines with lack of water to limit productivity, are quite impoverished. This contrasts with the lushness of moist tropical forests and coral reefs, where the abundance and diversity of species almost defy comprehension.

29. **Wetlands.** This lovely American egret makes its home in the Everglades National Park, Florida. Egrets live in marshes, swamps and estuaries, which are among Earth's most productive eco-systems. These wetlands are important providers of ecosystem services, especially in the cycling of nutrients. They also protect shore areas against severe storms, function as storage areas for excess water, thereby alleviating flood problems, and as natural reservoirs in time of drought. Wetlands more-over serve as nurseries for many important fish species and as way stations for migrating wildfowl, while harbouring an abundance of permanent plant and animal residents.

Human beings, like any other animal, are embedded in the planet-wide ecosystem – the ecosphere – which produced and still sustains civilisation. But now that expanding civilisation, like an approaching asteroid, threatens to cause an extinction episode unprecedented in 65 thousand millennia – and to destroy itself in the process.

CHAPTER THREE

The Shadow of Humanity

For more than four billion years, Earth got along without human beings. After the demise of the dinosaurs some 65 million years ago, the previously obscure mammals proliferated to replace them as the dominant animals on land. But even in the history of mammals, humankind was a latecomer on the evolutionary scene. The first known creatures that could reasonably be called human – short, erect, small-brained australopithicines – appeared less than four million years ago, although divergence of hominids from the ancestors of modern apes probably took place earlier. Human beings that were unmistakably modern *Homo sapiens*, however, go back no more than a few hundred thousand years.

Today something entirely unprecedented is happening. An element of the biosphere itself, an evolutionary newcomer, *Homo sapiens*, is threatening the entire living world, including itself. Evolution appears to have sown the seeds of the next giant extinction event. The

30. **Young lionesses learning to hunt.** The transmission of information from parents to young is a fairly common trait among non-human mammals and represents a simple form of culture.

beginning of the current episode of extinctions may be most conveniently dated as the time, about 10,000 years ago, when some groups of *Homo sapiens* gave up hunting and gathering and settled down to practise agriculture. But long before the advent of agriculture, there were important advances in the evolution of our species that made farming possible and all that followed.

The earliest known hominids already possessed certain distinctly human characteristics – especially the habit of walking on two legs. Another trait, shared to a degree with chimps and other apes but carried further by human species, was a relatively long life-span, accompanied by a prolonged infancy and childhood dependence on parents. Thus the trend was towards bearing one offspring at a time at intervals of several years, with parents investing more and more effort in rearing each one.

Also like most other primates, hominids were decidedly social animals. Unlike most of the apes, however, human beings usually form close, long-term pair-bonds (that is, they tend to have a monogamous mating pattern), and males are usually active participants in child-rearing. Moreover, the human sexual pattern is unique. There are no well-defined periods of oestrus ('heat') in human females; their sexual receptivity is essentially continuous, a circumstance that some believe adds strength to the pair-bond – encouraging the male to stay around and help rear the helpless infant. The development of the nuclear family may have arisen quite early in human evolution. Indeed, it has been suggested that the need for foraging males to carry food to dependent females and offspring might have been the impetus for walking on two legs.

But all this is highly speculative, and there are other speculations. As anthropologist Donald Symons points out, there is little reason to believe that continuous sexual receptivity of the female is responsible for the tight pair-bond. Gibbons are monogamous, but the female is not continuously receptive; in many human societies there is evidence that marriages may prosper in spite of rather than because of sexual obligations. Symons suggests that year-round sexuality may have evolved for other reasons.

For instance, being continuously receptive may have helped females with dependent young to beg meat from successful male foragers or hunters. Or perhaps concealing the female's fertile period made it more difficult for the male to prevent his mate from being inseminated by another. That is, lack of external signs of oestrus may have been a device evolved in human females to make it easier for the wife of a powerful (or rich) old man to have a child by a vigorous young man. If wives were 'in heat' for only a few days a year, guarding them for the vulnerable period would be much simpler for husbands.

But whatever the role of loss of oestrus in its evolution, the strong nuclear family may also have contributed to the evolution of the most striking constellation of human traits — a large brain, speech, and the transmission of cultural information between generations.

All organisms transmit information from generation to generation in their genes; that is how the offspring of a pine tree knows to grow into a pine, and that of a coyote into a coyote. Millions of years ago, though, our ancestors began to transmit significant amounts of information extra-genetically — by parents teaching offspring and by general sharing of knowledge among members of groups.

Such cultural transmission is not limited to human beings and their forbears; many large carnivores, for example, teach their young to hunt. But in the evolutionary lineage leading to our species, perhaps to compensate for a lower reproductive rate, greater intelligence, cooperation and the communication of non-genetic information became increasingly important for obtaining food and other resources, evading predators and other environmental hazards, and protecting the young. Thus, through natural selection, hominid brains gradually enlarged to accommodate the need to store and process the additional knowledge.

31. **Aborigine child learning to hunt.** A few isolated groups of people still survive by hunting and gathering, including these Australian aborigines near Liverpool River, Arnhem Land. This boy's spear tip is wrapped with paper bark so that he cannot injure himself or other children while practising. The adult's face is painted for ceremonies taking place at the time.

These changes came slowly at first. The fully upright australo-pithicines of the era of 'Lucy' (a famous recently discovered fossil over three million years old), had brain volumes of about 450 cc, roughly the same as modern chimpanzees. As recently as one and a half million years ago, the brains of later hominids had enlarged to only about 850 cc – slightly over half that of *Homo sapiens*.

The principal evidence of cultural advance in prehistoric human groups is in the increasing sophistication of the tools and weapons they used. Other animals, especially chimps, are known to fashion and use tools, but they do so relatively casually. The dependence on such implements for survival is characteristic only of human beings. Yet it was not until about two and a half million years ago that human tool-making reached the stage of development where recognisable tools can be found in the fossil record.

Throughout the millennia of hominid evolution, the cultural information that was transmitted doubtless consisted primarily of techniques for obtaining and preparing food, presumably including the use of sticks, stones and bones as weapons and tools, and the use of leaves and skins as containers; perhaps methods of starting, maintaining and using fire; warnings about hazards in the environment; rules for social conduct; and myths, traditions and rituals of the group. All this information was transmitted entirely by verbal instruction and demonstration.

Today, with brain volumes of around 1500–1600 cc, human beings must deal with an enormously more complex cultural environment than did their ancestors. Modern culture, while still dealing with the myths, traditions and rituals, encompasses such diverse new topics as science, international relations, gourmet cooking, terrorism, literature, art, music, automobiles, aircraft and nuclear warfare. And the means of cultural storage and transmission have broadened from brains, tongues and hands to books, films, laser discs, video and computer tapes, telephones, radios and television sets. One result is that in barely two million years, one kind of organism has created an astonishingly new and successful way of dealing with environmental change – a way that to a large degree circumvents the need for rapid genetic evolution.

The cultural revolution was undoubtedly an important element in this success, especially in the last hundred thousand years or so. Following the appearance of *Homo sapiens*, tool-making became an art, and hunting of large animals also became a regular practice, supplanting and supplementing the earlier dependence mainly on plant foods, small animals and possibly scavenged kills of other predators. As they perfected their hunting skills, the new hominids flourished and spread to occupy all the major land areas of the planet by the time the last ice-age glaciers were in full retreat, 10,000 years ago. It may have been their very success as hunter-gatherers that eventually gave rise to the agricultural revolution.

Humanity had a significant impact on the biosphere long before the advent of farming, however. It has been proposed that much of the savanna of tropical Africa was created by the burning of forests and woodlands by human hunters many thousands of years ago. But an even greater effect was the hunters' contribution to the extinction of most of the so-called 'Pleistocene megafauna' – herds of mammoths, woolly rhinos, giant sloths, massive bison and many other impressive herbivorous animals that roamed ice-age landscapes and were preyed upon by sabre-toothed cats, dire wolves and other carnivores.

32. **Mammoths and human hunters.** During the heyday of the last glacial epoch, perhaps 35,000 years ago, human hunters may have been skilful enough to tackle giant mammoths, relatives of today's elephants. Indeed, our ancestors may have been partly responsible for the disappearance of these huge creatures.

The disappearance of these great animals from the Old World – Europe, Asia and Africa – was gradual, suggesting a slowly improving human hunting technique. But in the New World, they vanished with surprising rapidity – and suspiciously soon after the first human groups, who by then were skilful hunters, invaded North America. With the spread of savannas and the demise of these prehistoric beasts, the shadow of humanity first began to creep across the planet.

In retrospect, the agricultural revolution may prove to be the greatest mistake that ever occurred in the biosphere – a mistake not just for *Homo sapiens*, but for the integrity of all ecosystems. By taking up agriculture, people began to alter the biosphere in ways that in the long run would prove much more far-reaching than the removal of the giant herbivores and just as irreversible.

A contemporary observer could not have perceived the destructive potential of small groups of people settling in one place to encourage the growth (and control the evolution) of a few species of plants that

59

would become known as crops. The enterprise was small in scale and not always successful. Unfavourable weather could reduce crop yields catastrophically. And other plants, undesirable from the human viewpoint (weeds), could choke out the crops in spite of the farmers' best efforts. Diseases of the crops could sweep through a farming area, ruining a year's harvest; or other organisms, sharing human tastes, would devour part or all of the crops in the field or in storage. Pests have always been part of farming.

But despite weather, weeds, diseases and pests, the farming populations expanded and spread. While the agricultural way of life was not entirely secure, it did allow many times more food for human beings to be produced in an area of a given size than could be extracted from natural systems by hunter-gatherers.

33. **Eskimo hunter.** Among the few remaining peoples that still hunt and gather are some of the Eskimos. Hunting is a serious business and is learned early; this boy is bringing home a brace of ptarmigan. Unfortunately, Eskimos are increasingly being integrated into industrial society, and their legendary skills in hunting and survival are rapidly being lost.

By the time large-scale civilisations rose and flourished, beginning four or five thousand years ago, extensive tracts of natural systems in well-watered temperate and sub-tropical regions in much of Europe and Asia had been replaced by human-directed systems, sometimes called agroecosystems. Within the last two millennia, large portions of Africa and the Americas have similarly been converted to agriculture. Always, the introduction of farming has been followed by an enormous, if usually slow, expansion of the human population it supported.

Farming slowly took over warm areas of the world where rainfall (or opportunities for irrigation) permitted. But human groups in cooler and drier areas often became dependent on migrating herds of large herbivores – such as wild horses, cattle, sheep, bison and caribou – and sometimes gradually assumed control of herds while protecting them against other predators. So probably began the domestication of large animals. Some of the early nomadic herding cultures have persisted to the present day, such as the Tuareg of the Sahel.

The beginning of agriculture led to the development of towns and cities, since farmers soon could provide food for more than just their

own families. Urbanisation in turn led to an explosion of cultural evolution, and eventually to the science and technology that now both support and threaten our lives. Early on, it also began to produce the kind of political structure that has plagued Earth ever since – a structure of nation states and their inevitable companion, war. That organised warfare existed at least 7,000 years ago, only a few millennia after the origins of agriculture, is attested to by the heavy fortifications of Jericho at that time.

34. Nomadic herding. These Samburu near Ngurunit in Kenya, still follow an ancient way of life tracing to the first domestication of animals thousands of years ago. They migrate with their cattle in search of productive pasture.

In agriculture also were planted the seeds of runaway human population growth. For the farming groups, the settled agrarian lifestyle seems to have stimulated that growth in several ways. Demographers long thought that the increased security of the food supply might have led to lower death rates in farming groups. Anthropologists now think that the food security was not that great and that it may have been largely offset by increased diseases among the larger and denser farming populations. Birth rates, on the other hand, might have risen when it was no longer necessary for mothers to carry each child until it was old enough to keep up with the roaming hunter-gatherer group. The availability in farming communities of softer foods also could have led to earlier weaning and thus more frequent child-bearing.

Whatever the reason, the global human population, estimated to be about five million at the beginning of the agricultural revolution around 8000 BC, reached a size of 500 million around 1650 AD. During that period, the population thus doubled in size roughly every 1,500 years. This increase was very slow and not at all steady. The overall average growth rate conceals minor accelerations in good times and numerous setbacks, such as the Black Plague, which decimated European populations during the late middle ages.

Between 1650 and 1850, the human population doubled again to 1,000 million (one billion) people. This surge in population growth was concentrated in the West, where it was stimulated in large part by the breakdown of feudalism and the opening of the New World to exploitation by Europe (and the acquisition from America of two new staple foods, maize and potatoes). These changes were accompanied by advances in farming systems, such as the use of clover (which, like other legumes, has mutualistic nitrogen-fixing bacteria living in its roots) as 'green manure'. A similar spurt in population growth occurred in Asia, especially China, where it was perhaps due to a long period of peace, stable agricultural policies and the introduction of new crops, maize and groundnuts (peanuts).

This period also ushered in the industrial revolution – the beginning of manufacturing in an organised way outside homes and the increasing use of machinery. Like agriculture, the industrial revolution marked an enormous change in humanity's impact on the planet. The use of energy, other than humanpower or horsepower, brought far-reaching changes. Wood, peat and coal had long been used for heating buildings, cooking and metal-working. Now, with the invention of the steam engine, the energy of heat could be harnessed to run machines, and the rate of exploitation of Earth's vast store of fossil energy surged upward.

35. A water wheel. Power for manufacturing in the early industrial revolution in England was provided by water wheels like this one illustrated around 1860. Falling water from lakes behind dams is still an important source of energy, but it is used today to spin turbines that run electric generators.

36. Bradford in the 1860s. The town of Bradford in Yorkshire, near the heart of a coal-mining region, was an expanding factory town in the mid-nineteenth century when the Industrial Revolution was in full swing in England. In those days, such clouds of smoke were viewed as signs of progress, and their possible health impacts were little-known and generally ignored.

37. **Steam engine.** Built in 1866, this James Watt beam engine in the Crossness Sewage Works pumped sewage from the London main sewers to ground level. Originally, the sewage was pumped directly into the Thames; later it was sent to settling tanks. The pump could lift 100 tons of sewage per minute. The harnessing of inanimate energy thus helped to improve human health – in this case by improving sanitation – although it also created new hazards from air pollution and possible large-scale accidents.

In the late nineteenth century, the convenience of petroleum as a fuel was discovered; within a generation it revolutionised transport, agriculture and almost every other aspect of human life. With a cheap and seemingly abundant source of energy, mineral and fossil fuel resources could be mobilised at a much faster rate, and considerably more food could be produced on a given piece of land. Humanity had discovered a set of previously untapped resources, unavailable to any other animal, which allowed more complete exploitation of older resources. As a result of these advances, many more people could be supported.

38. **Horse-drawn combine.** The era of the Industrial Revolution also brought the occupation of the previously uncultivated lands of North America and the beginning of modern, large-scale farming. This combine for harvesting wheat in the Great Plains was photographed in 1903. The pattern of using animals to power farm machinery, begun with the horse-drawn plough, continued as more complex devices were invented. This agricultural operation was thus still run largely on solar 'income'; the heavy contribution of fossil fuel 'capital' was still in the future.

With these changes, humanity began to shift its dependence away from the 'income' provided by solar energy in the form of food and other products of the biosphere. Instead, people became increasingly dependent upon the 'capital' portion of their inheritance, the non-renewable resources of Earth: minerals, stored groundwater and fossil fuels.

The shadow of humanity was growing longer and deeper. By the nineteenth century, not only had vast areas of land been taken over to produce food for human beings – at the expense of most of its other inhabitants – but entirely new ways of causing environmental damage had been invented and deployed. Even in the eighteenth and nineteenth centuries, air pollution could be a problem in large cities like London. So was pollution of rivers and streams, whether from sewage overload, mining or industrial effluents.

As with the earlier agricultural revolution, however, the profound implications for Earth's future of the industrial revolution were not perceived by the people of nineteenth-century Europe and North America. They only saw that life was becoming a little easier for themselves and the future somewhat brighter for their children.

39. **Opening the West.** This illustration from *Harper's Weekly* in 1875 shows the building of the first trans-continental railroad in the United States. Completion of the railroad was a major step in unifying the expanding young nation and in stimulating its economic growth. Railroads everywhere facilitated the exploitation of natural resources and led to the accelerated depletion of resources as diverse as timber (used for ties and firewood) and the American bison.

40. **An early oil field.** The discovery in the second half of the nineteenth century of the usefulness of petroleum swiftly transformed life-styles in the industrialising nations and especially revolutionised transportation. This oil field was near Taft, California, around the turn of the century.

Less than a century was needed to double the world population after 1850. By the time of our births in the early 1930s, there were two billion *Homo sapiens* on Earth. That dramatic surge could largely be traced first to reductions in death rates following improvements in simple sanitation (the provision of clean water, the use of soap, and the construction of sewer systems) and later to an organised assault on epidemic disease by an advancing medical technology. This acceleration in population growth, however, was centred in the industrialising western world and those parts of other cultures in close touch with it; most other societies as yet had experienced little change in death rates.

During the nineteenth century, birth rates had also begun to decline in western nations, largely in response to social changes related to industrialisation, especially changing attitudes towards children. As more and more people left farming, moved to cities and became involved in manufacturing, children were no longer viewed almost purely as assets, useful as labour on the family farm and serving as a sort of old-age insurance. Instead, they became rather expensive consumers, requiring housing, food and education. People responded to these higher costs, as well as to the children's increasing chance of survival, by limiting their families. This phenomenon – a decline in birth rates following the decline in death rates in industrialising societies – became known as the demographic transition.

In the first half of the twentieth century, as the demographic transition was advancing in western nations, it spread to the industrialising eastern European nations and the Soviet Union. But it was not until after World War II that the reductions in death rates that had occurred earlier in the West spread to non-industrialised parts of the world.

Then, technologies for disease control – especially DDT and other pesticides to attack disease-carrying mosquitoes, and antibiotics to control bacterial infections – were widely disseminated in Asia, Africa and Latin America. These easily applied substances suddenly brought considerable control over malaria and numerous infectious diseases, causing human death rates to plummet, especially among the very young. By the 1970s, mortality rates in many less developed countries were approaching those in industrialised nations.

This death-control technology was imposed from the outside, however; by and large it was not accompanied by industrialisation or other fundamental changes in the societies affected. As a result, there was no sign of a demographic transition; in many countries, there is still no evidence of one. Although death rates plunged, birth rates remained high. Since the rate of population growth is determined by the gap between the birth rate (input to the population) and the death rate (output), the consequence was an unprecedented population explosion in the less developed nations.

Population growth rates in developing nations today typically range between 1.5 per cent and 4 per cent a year, with most of them averaging between 2 and 3 per cent. With an annual growth rate of 3 per cent, a population will double its size in 23 years; at 4 per cent, it doubles in a mere 17 years.

Since the majority of the world's human beings live in the less developed nations (about three out of every four in the 1980s), the entire world population has also been increasing faster than ever before. Between 1930 and 1976, the global population doubled to four billion souls. During the early 1960s, the worldwide average growth rate was an estimated 2.2 per cent a year, fast enough to double the population in just 32 years. The human population has now passed the five billion mark. Thus, two and a half times as many people now live on Earth as existed when the two of us were born.

Even though the annual percentage rate of growth has lowered significantly since the high point a couple of decades ago, that percentage annual 'interest' (1.7 per cent) is being applied to an ever growing stock of human 'capital'. Consequently, each year sees a record absolute growth in the population. This year, the population will be swelled by more than 84 million additional people to feed, clothe, house, medicate, educate and employ – a population equivalent to a new United Kingdom plus Canada.

While the developing nations have been having their turn at the population explosion game, the demographic transition has continued in the industrialised countries. Some European nations, such as the UK and Sweden, have essentially stopped growing. They have about reached zero population growth (ZPG); that is, their birth rates are nearly equal to their death rates. Populations of a few countries – Austria, West and East Germany, Denmark and Hungary – have actually begun to shrink gradually in size, because their death rates are now slightly higher than their birth rates (outputs are greater than inputs). If the present fertility rate of the United Kingdom continues, the British population will soon embark on a slow decline in numbers.

In most of the industrialised nations, however, populations are still slowly expanding. In the United States, Japan, Canada and the Soviet Union, among others, rates of natural increase (the excess of births over deaths) are about 0.5 to 0.7 per cent a year. To put this into perspective, if those rates continued unchanged, the populations would double in 100 to 160 years.

But those rates are not likely to remain unchanged, for several reasons. First, reproductive behaviour in modern societies is by no means constant; it fluctuates in response to any number of factors such as economic conditions, wars, participation of women in the work force, and so forth. In addition, there are factors of population structure (the proportions of people of different ages) at play that will cause birth rates in these countries to decline further in the next few decades, even without a significant change in the average person's reproductive behaviour.

The United States, for example, reached a demographic landmark in the early 1970s, when its fertility dropped below 'replacement reproduction'. At the replacement level, each generation of parents just replaces itself in the next generation. With the mortality rates prevailing in modern industrial societies, replacement reproduction implies that each woman in the population will bear an average of 2.1 children in her lifetime, replacing herself and her partner. (The extra fraction compensates for those children who die before reaching reproductive age.) If replacement reproduction is exactly maintained for long enough, the population will eventually stop growing and remain stationary indefinitely.

Why doesn't it stop growing immediately? The reason is that rapidly growing populations, with high birth rates and low death rates, have a very large proportion of young people, who will mature and have children and grandchildren themselves before they reach old age and begin to die in large numbers. In many developing nations, for instance, 40 to 50 per cent of the population are under the age of fifteen; they are the parents of the next generation. Even if the current

41. An Edwardian family. By the early twentieth century, in industrialising nations of the West, most children survived their early years. This picture shows the Terry family of Greenwich, in 1914, with sixteen of their nineteen children. Some couples were beginning to practise birth control by then, but large families were still very common.

42. A contemporary family in Brazil. In many developing nations, mortality rates among children have fallen very low, but few people have responded by limiting their families. In 1977, this 30-year-old widow was left with eight children, the two youngest of whom are partially paralysed, and a ninth on the way. She lives in Rio Grande de Norte, in north-eastern Brazil, a region of deep poverty, in part because of ecological mismanagement, especially deforestation. In Brazil, as in many other Latin American countries, the poor have little access to family planning information or materials, although members of Brazil's wealthy and middle classes now almost universally limit their families.

crop of youngsters produces an average of two children per family, the population will continue to grow throughout their lifetime, just because there are so few people in the older generations who will soon die. Because of this momentum built into the structure of the population, once replacement reproduction is reached, it takes about a lifetime – 65 or 70 years – for a previously expanding population to stop growing.

This situation is extreme in less developed countries, where death rates have fallen precipitously while birth rates generally have remained high, vastly swelling the ranks of young people. But in many industrialised countries, especially the United States, the Soviet Union and several former British colonies, the same phenomenon exists because of the post-World War II 'baby boom'. Between 1945 and 1970, birth rates were moderately high in these countries, and accordingly growth rates were 1 per cent a year or more. The baby-boomers of the United States are now entering middle age; even though their fertility is low, they have caused a small 'boomlet', simply because there are so many baby-boomers all having *their* children at once.

Since 1973, the average American family size has been somewhat below replacement level – small enough so that if fertility (the average number of births per woman of reproductive age) did not rise again, the US crude birth rate (which is measured not relative to reproductive women but to the entire population) eventually would slowly decline. Meanwhile, the crude death rate (also measured relative to the total population) would rise until it met the falling birth rate. Both changes would be due to an increase in the average age of the population. When the birth and death rates were equal, natural increase would cease.

One other factor can change population growth rates: migration. Even with birth and death rates in balance, a population would continue to grow if the numbers of immigrants exceeded those of emigrants; inputs from all sources must balance outputs for ZPG to be achieved. In the United States today, without immigration, the present low reproductive rate would bring an end to population growth within a few decades, after which there would be a slow population decline.

But that will not happen as long as the United States has a substantially higher rate of immigration than emigration (inputs higher than outputs). Probably over a million immigrants per year enter the country, many of them illegally (and therefore uncounted). Added to natural increase, this influx of people boosts the current annual US population growth rate to over 1 per cent. A continuation of immigration at this level, or an increase, would appreciably delay the end of population growth and swell the ultimate peak population size.

Indeed, a net addition of a million or more immigrants per year would postpone the end of growth indefinitely, unless there were a further drop in fertility.

Clearly, small changes in rates can make a vast difference later on. If there were no immigration after the mid-1980s, and fertility remained the same (about 1.8 children per woman), the US population would reach its peak size around 2030 at less than 250 million, then decline to around 200 million fifty years later. By contrast, if fertility rose to slightly above replacement to an average of 2.2 children per family and net immigration were 1 million per year, the US population would soar past 510 million by 2080, more than doubling in less than a century, and would continue growing until there was a change in one or both vital rates (birth or death) and/or migration.

Of course, the people added by immigration to the United States (or any) population are subtracted from somewhere else. Today the great majority of international migrants move to more industrialised nations from less developed nations, thereby somewhat dampening the rapid population growth of the latter. But that dampening is relatively inconsequential in a global context.

Most developing nations are a long way from replacement reproduction, and their populations are almost inevitably destined to expand considerably before their birth rates can be reduced that far. China and a handful of small nations, however, are within striking distance, if not already there, and their fertility will probably have fallen below replacement well before this century is out. Even so, the built-in momentum from the earlier high birth rates will sweep their populations to much larger sizes before natural increase ceases, because of the preponderance of young people – unless there is a substantial change in death rates.

With no rise in death rates, the populations of most developing countries would continue to expand by 30 to 100 per cent after reaching replacement fertility, depending in part on how rapidly and how far their birth rates fell. And, for the majority of developing nations that have yet even to approach that level of fertility, demographic projections anticipate a doubling or tripling of populations before growth finally ends more than a century from now. Thus, carried mainly by the momentum inherent in the population structures of developing nations, the global population is projected to reach a maximum size of approximately 10 billion (give or take a billion or two) – about twice its present size – before growth ceases.

Since the 1950s, the populations of less developed nations have been expanding rapidly at a remarkably constant average rate (if China is excluded) of about 2.4 or 2.5 per cent a year. For a long time, social scientists complacently assumed that the process of industrialisation

71

would automatically bring about a demographic transition, as had happened earlier in the West. But, to their surprise and dismay, no demographic transition had appeared by the 1970s.

Indeed, although the economies of the developing nations were expanding at historically unprecedented rates — on average, even faster than the also rapidly growing economies of industrial nations — the economic gains were quickly eaten up by population growth. If an economy was expanding by 3.5 per cent a year, and the population by 3 per cent, the net economic growth per person in the population was only 0.5 per cent. In the industrialised world, by contrast, owing to the much slower population growth, economic gains per person were appreciably greater throughout the postwar period, despite a lower rate of economic growth in most industrial nations.

This disparity contributed to a steadily widening economic gap between industrialised and non-industrialised nations. The average individual in developed regions was gaining in wealth much faster than the average person in less developed nations; the already rich were getting constantly richer, while most of the poor were scarcely making headway at all. Indeed, especially since 1972, a good many of the poor have fallen tragically far behind.

The character and meaning of the widening income gap (now almost a chasm) between rich and poor nations is perhaps most starkly revealed in the world food situation — which itself has been transformed since World War II. In the last four decades, worldwide food production has risen substantially — as, of course, it had to in order to feed a population twice as large. Production of grains — which comprise the basic source of nutrition for human beings — has more than doubled in that time. Thus, on average, there has been a small annual increase in food produced per person in the population.

Unfortunately, the increases in food production have not been evenly distributed. Like the growth in national economies, increases in food production in poor countries have generally been faster than in rich countries, but the high rate of population growth has consumed most of the gain; increases per capita have been disappointingly small. While the abundance of food supplies in rich countries has become almost embarrassing (and an economic problem of no small magnitude), substantial portions of the populations of most poor countries still have too little to eat. Even in some rich countries, such as the United States, hunger has not been entirely banished.

Since 1972, in many of the poorest nations (which also have some of the fastest growing populations), principally in Africa south of the Sahara, the amount of food produced per person has been steadily dropping. Even in 1972, these populations were seriously underfed. While imported food might make up the gap between needs and

43. Modern agriculture. These huge diesel combines are harvesting wheat in eastern Washington state in the United States, the world's leading grain-producing nation. The industrialisation of agriculture, as exemplified by this machinery with its dependence on fossil fuel 'capital', has contributed to substantial increases in food production in the West since World War II. Ecologist Howard Odum once remarked that 'industrial man no longer eats potatoes made from solar energy; now he eats potatoes partly made of oil'.

44. Supermarket. A symbol of affluence and of the abundant food supply in developed nations is this supermarket in north London. Fossil fuels not only make enormous contributions to the growing and harvesting of the foods shown here, but also to their processing, distribution, marketing and handling at home or in restaurants. More calories of energy go into producing the canned goods here than they yield in food energy.

73

supplies, these countries are often too poor to afford massive imports, so their already hungry populations go even hungrier.

In spite of heroic increases in global food production over the last several decades, a large proportion of the world's people are still poor and hungry. Even though that proportion has been reduced somewhat, the absolute number of hungry human beings is much higher now than in the late 1940s, just because the total population has nearly doubled. The United Nations Children's Fund (UNICEF) estimates that some 15 million infants and small children die unnecessarily each year from malnutrition and other poverty-related causes.

A profound shift has also taken place in the pattern of international trade in food. Before 1940, nearly all the world's regions exported food to other nations. The only important exception was Europe, which imported more food than it exported. Today, western Europe has joined the dwindling list of regions that export more food than they import, whereas the vast majority of nations are now dependent to some degree – and sometimes heavily so – on imported food. Only a handful of countries (primarily the United States, plus Canada, Australia, New Zealand, Argentina, Thailand and the European Community) supply the international grain market for all the rest, including the Soviet Union, eastern Europe, Japan and most of the poor countries.

In addition, until the late 1970s, the primary importers of food were rich countries, which could easily pay for the imports by exporting manufactured goods. In the last decade, though, rapid population growth in the poor countries increasingly outstripped their ability to keep raising food production. Less and less virgin land remained available to open for agriculture, and the scope for increasing productivity on existing farmland (the famous green revolution) was more and more constrained by both biological limitations and the economic realities of poor countries.

In 1974, the oil crisis jolted the world into awareness that petroleum, as well as other fossil fuels, was indeed a finite resource. The OPEC cartel quadrupled the price of oil on the world market, sending shock waves through the global economy. The rich countries were forced to adjust to much higher energy costs, and as a result their economies suffered inflation, a slowing of growth and rising unemployment. A further doubling of oil prices in 1979 caused even more economic dislocation and contributed to a severe, worldwide recession.

On the positive side, though, the higher costs of oil and natural gas induced the rich countries to institute measures to conserve these commodities; the amount of fuel that was wasted or used inefficiently was markedly reduced. A beneficial side-effect of the reduced

45. **A slum in Lagos.** In contrast to the affluence of the industrialised nations, substantial numbers of people in many developing nations are extremely poor and often hungry. This is a 'street' in Lagos, the capital of Nigeria. There are no services in this neighbourhood for water supply, public transport, or sewage or waste disposal. In Nigeria, a child is more than ten times as likely to die before its first birthday as is a child born in Great Britain or the United States.

consumption of energy was to lower the environmental impacts of its use.

But a heavy penalty for the higher prices was paid in the poor countries, whose modernising economies, and especially the agricultural sectors, were set back by the suddenly unaffordable fuel and other petroleum-based products such as fertilisers and pesticides.

Much of the momentum of development was lost, and food production increases faltered in the late 1970s. Many developing nations went deeply into debt, trying to pay huge oil import bills.

With their fast growing populations, struggling economies and faltering food production, many poor nations are facing catastrophe unless dramatic steps are taken. And there is very little sign of such steps materialising, even though the need for them has been increasingly evident for several decades. The poorest of these nations – ones like Chad – are not 'less developed countries', as they euphemistically prefer to be called. Under the current global regime, they would best be considered 'never-to-be-developed' countries, nations that have been written off by arithmetic, if not by intent. The shadow of humanity has so shrouded their landscapes that they seem condemned to perpetual darkness.

The deepening shadows are not restricted to the poorest nations, however; they are merely the most vulnerable. As could be expected, the stresses on the global system have led to collapses in the weakest areas, as are described in the next chapter. The enormous, widening gulf between rich and poor peoples, as illustrated by the demographic differences and the food situation, pervades almost every aspect of the human predicament. It colours global politics, causing jittery governments to pile up deadly, expensive and unusable weaponry to protect their land and access to resources, and thus undermines the security of all nations. And the rich/poor division seriously hinders any attempts to address the deepening predicament. Yet that predicament – the overpopulating of Earth by humanity and the squandering of its unique inheritance – is a cause of that same tragic and perilous imbalance.

By the 1980s, the high costs of overpopulation and the heightening dependence of human beings on non-renewable resources had begun to be felt. Staying in the race to produce enough food for more and more people is becoming increasingly difficult. Economies seem to be losing the battle to grow faster than the populations they sustain. Less and less 'virgin' land is available to take over for human use, and natural ecosystems are in full retreat. Earth and its living passengers are taking a beating the likes of which has not been seen in at least 65 million years.

As our species is forced by its own actions to burn its capital in earnest, however, the consequences are borne not only by the other life-forms on this lonely space vehicle, but they also fall heavily on our fellow human beings. That lengthening shadow not only darkens the lives of our contemporaries but the prospects of future generations as well.

PART TWO

Earth's Predicament

CHAPTER FOUR

The Costs of Numbers

To the rich, the poor are often invisible. The massive suffering of almost half of humanity does not often come to the attention of the affluent quarter. But when rail-thin, fly-covered, naked, dying children appear on the evening news, the true state of the world can intrude into the living rooms and consciences of those who have never had to miss a meal or go without decent shelter or clothing.

People in the rich countries have been treated in recent years to poignant television and newspaper reports about the starving victims of a widespread severe drought afflicting the southern fringe of the Sahara desert all across the African continent – the Sahel. The causes of the drought itself are not entirely clear, although small changes in the general circulation pattern of the atmosphere can alter the strength

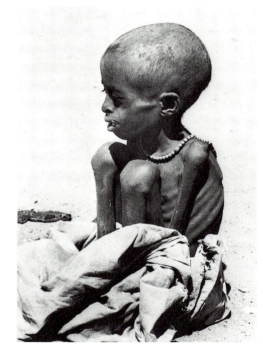

46. **Starvation in the Sudan.** While people in the rich developed nations have too much to eat, those in poor countries often have too little. This child of the Hadendawa people is a victim not only of poverty but of a famine caused by overexploitation of the area's fragile arid environment combined with an intense drought that afflicted much of Africa in 1984 and 1985. The Hadendawa are facing virtual extinction as a culture because of the desolation of their land.

79

47. Desertified land. Cattle, goats, sheep and camels take water at El Beshiri oasis, Sudan. The herds have trampled all the vegetation as far as the eye can see around this important water-hole. The severe drought in the mid-1980s accentuated the consequences of desertification caused by overpopulation of people and their domestic animals.

of the monsoons that bring rain to the region; when the monsoons are weak or delayed, the rainfall is reduced. Historical evidence suggests that prolonged drought – a sequence of years with abnormally low rainfall lasting two decades or more – is a recurring phenomenon of the region. A drought similar to that of the 1970s and 1980s seems to have occurred in the early nineteenth century, but there is little evidence that it produced a great famine.

What is new, then, is the tragic suffering of the people living in the Sahel, and the intrusion of that suffering into the world view of the well off. Profound changes in lifestyle resulting from modernisation processes, political changes, and, above all, a dramatic increase in the populations of the resident people and their domestic animals, have greatly increased the vulnerability of the people to a change in climate. In the past, the sizes of the herds were limited by the available water. In the 1960s, numerous tube wells were installed to provide abundant water, the herds expanded, and soon the delicate desert and savanna grasses began to be overgrazed. The consequent process of deterioration of soil and vegetation is known as desertification.

The drought first appeared in the late 1960s, producing a regional emergency in 1974. Coordinated emergency relief that year saved millions from disaster. Then the drought abated slightly, and the outside world forgot about the Sahel – until 1984, when the ongoing crisis was suddenly rediscovered. This time, most of the attention of the West focussed on Ethiopia because political points could be scored on its Marxist government, whose policies toward two rebellious provinces had greatly exacerbated the effects of the drought. But the famine was by no means limited to one nation. No fewer than 20 African countries suffered disastrous harvests in 1983 and 1984, and tens of millions of people were facing starvation. This time the trouble extended to much of southern Africa as well as the Sahel.

Once the news of the famine reached the outside world, however, the response was immediate and generous. People gave money for emergency food to be sent to Ethiopia and the other hungry countries; rock musicians in the United States, Canada and the United Kingdom jointly organised a benefit performance, which collected many millions of dollars; and the international relief agencies also mobilised to supply food and assistance.

By 1986, the emergency seemed to be over; rain had returned to the Sahel and harvests were better. People – and journalists – in the developed nations turned their attention to other matters. The emergency may be over, for now, but the problem is a long way from being solved. The famine was a symptom of an underlying disease, thrown into stark relief by the chance occurrence of a change in climate. Just as an individual with a bad cold may find his symptoms

48. **Ethiopia's military might.** The Marxist government receives most of its arms from the Soviet Union, and uses them against its own rebellious people in the northern provinces. When famine struck those provinces, the Ethiopians pursued the war rather than provide food for the hungry, at the same time shipping grain to the Soviet Union to pay for the arms. These weapons were paraded in Addis Ababa in 1984 to celebrate the tenth anniversary of the Ethiopian regime.

intensified by exposure to harsh weather, so the weaknesses of the overburdened ecosystems on which the people of the Sahel depend were intensified by the drought. The combination of very rapid population growth, a breakdown of traditional land-use practices and substantial mismanagement of the fragile terrestrial ecosystem, plus bad luck with the weather, brought disaster. The result was essentially a collapse of the systems, from which recovery will be extremely difficult and slow, if it is possible at all.

The underlying ailment is human overpopulation, and the Sahel is simply one place where an acute manifestation of it was imposed on a highly vulnerable region. As we will see, overpopulation can have a terrible impact on our living companions on the planet and, through them, on life support systems and thus on *Homo sapiens*. But overpopulation can have directly unpleasant effects on human beings and their societies as well. The tragedy of the Sahel is not an isolated situation; many developing areas seem doomed to follow a similar path.

The goal of most developing nations, naturally, is to achieve the affluence of the rich countries. People in most poor nations have rising expectations; sadly, though, they have plummeting prospects. Overpopulation, made ever worse by continuing rapid population growth, is a great barrier to sound development in many poor nations. The urgent requirement is always to provide for the people in need today; it is too soon to worry about providing for the even greater numbers of a decade hence. Unfortunately, the process of meeting today's needs all too often diminishes the resources and opportunities for the future.

Consider, for example, the plight of one developing country: Kenya, in many ways the jewel of Africa. Kenya is blessed with cool highlands, some fine farmland, spectacular scenery, magnificent flora and fauna and enterprising people. Kenya could have a secure future with an economy based on tourism and agriculture; along with some other eastern and southern African nations, it is one of the few remaining places on Earth where even a glimpse of the nearly vanished Pleistocene megafauna is still possible. The large animals of Kenya's game parks are a priceless part of the human heritage.

But that rich heritage is swiftly vanishing from Kenya under the impact of a population growth rate of 4.1 per cent a year, the fastest rate of natural increase ever recorded for a nation. Kenya and many other African countries have extremely high growth rates, partly because Africa was the last continent to feel the impact of public-health technologies exported from the West. Death rates, especially infant and child mortality rates, have declined – as they did in all less developed nations after the introduction of DDT and penicillin. But there has been no sign of a demographic transition in Africa's tribal societies, no compensatory drop in birth rates – in fact, just the opposite. Healthier mothers are having *more* babies, not fewer.

Kenya has been among the most successful of tropical African nations in reducing its infant mortality, although it is still six or seven times as high as in most rich nations. Meanwhile its birth rate has soared. Consequently, the annual birth rate is now about 54 per thousand people in the population, and the death rate is only about 13 per thousand. Because of Kenya's rapid natural increase, the age composition of the population is heavily weighted toward the young: slightly over 50 per cent of Kenyans are under 15 years of age, and only 2 per cent are 65 or over. In comparison, the proportion of young people under 15 in the almost stationary populations of Europe is no more than 22 per cent, whereas that of the elderly ranges from 12 per cent upwards. If the current growth rate is maintained, the population of Kenya will double in the next 17 years.

What do these demographic statistics mean for Kenya's future? Above all, they mean that for Kenya simply to maintain today's inadequate standard of living, the nation in essence must double every amenity for the support of human life within 17 years. The mid-1980s stock of homes, schools, stores, factories and hospitals will have to be duplicated in about the time it takes for an individual to go from birth to finishing secondary school. The capacities of road, rail, port and airline facilities today will have to be doubled by shortly after the turn of the century. The productivity of the agricultural, forestry and water systems will have to be doubled by then, too, as will the presently inadequate stock of physicians, nurses, teachers, engineers and so on.

What are the chances that Kenya can accomplish this? It would be a virtually impossible task for the United States to duplicate its physical and trained human capital in a matter of 17 (or even 30) years. Yet the United States is rich, most of its people are literate and speak the same language, and it has a wealth of indigenous natural resources, as well as the economic, political and military clout to acquire other resources from the far corners of the globe.

Kenya has none of these advantages. It is a conglomeration of tribes with diverse cultures and languages and a history of mutual hostility. Its adult literacy rate, one of the highest among African nations, is still only about 50 per cent, although primary school enrolments are now approaching 100 per cent. No more than 10 per cent of Kenya's land is considered potentially arable, as compared to over 25 per cent in the United States and 40 per cent in western Europe. Kenya is already running out of adequate farmland; farmers are pushing more and more into fragile semi-arid regions and exerting pressure on the irreplaceable national parks.

49. Desertification in
Kenya. Extremely rapid
population growth has led
to overgrazing in Kenya's
arid northern region, which
borders the Sahel. The roots
of this acacia tree near Korr
have been exposed as the
soil has eroded away.

With its enormous proportion of young people, Kenya faces a critical
unemployment problem in the near future, unless a massive increase
in productive jobs can be generated in the agricultural sector. In 1984,
the World Bank estimated that the labour force in wage employment
(only about 15 per cent at present) would increase by two and a half
times in the last quarter of the twentieth century, and that a similar
proportional increase would have to be absorbed by the already
faltering agricultural sector.

In Kenya today, each farmer's land must be divided after his death
among an average of four sons. In many areas, such subdivision has
already reduced the average size of landholdings far below what can be
efficiently farmed by one family. Much of the best land, moreover, is
dedicated to growing cash crops for export, further intensifying the
pressure on the land used for domestic food crops.

Of course, if Kenya tries to improve the productivity of its
agriculture by mechanisation, fewer, not more, farm workers will be
needed in the future. Without modernisation, though, the agricultural
sector will not only be unable to provide jobs for the fast-growing
army of the unemployed, it will have little chance of meeting the need
for food production increases of over 4 per cent a year, just to keep
feeding the population at the present inadequate level.

For some time now, rural poverty and the land squeeze have been
driving people into the cities, principally Nairobi, whose population is
growing at about 8 per cent a year, roughly twice the rate for Kenya as
a whole. Jobs there are also far too few, but people find ways of
making a living, sometimes legal, sometimes not. Cities in poor

50. **A shanty town in Nairobi.** Such self-built communities are commonly found on the fringes of large cities in developing countries. In Kenya, as in many other African countries, young men migrate to the cities seeking work, while their families often remain behind in the countryside.

countries are typically ringed with self-built shanty towns in which the urban migrants live, often without clean water supplies, sanitation or other amenities. Nairobi is no exception.

Frequently, the governments of poor nations prefer to keep food prices artificially low in cities, ignoring the needs of the agricultural sector. It is in cities that riots and revolutions are usually sparked, and discontented urban masses spell trouble for politicians who wish to remain in power. The result is to depress the agricultural sector further, reducing any incentive for farmers to increase harvests and accelerating the flight of unemployed rural workers to the city. One might expect this route to be followed in Kenya; Nairobi will continue to be overwhelmed by mass migration from the countryside as rural unemployment and displacement from the land reach unmanageable levels.

In recent years, Kenya's agricultural production has fallen behind population growth, causing a 15 per cent decline in annual food production per person since 1972 and a massive increase in food imports, to the detriment of the country's balance of payments. Even though Kenya is one of the better-off nations in East Africa, a significant proportion of its people have too little to eat. In the mid-1980s, food production in Kenya was rising by an average of barely 1 per cent a year, partly because the country was affected by the drought in the Sahel.

Sooner or later, if current trends continue, pictures of starving Kenyan peasants will be shown on the TV screens of the rich countries – as were the victims of the Sahel crisis. If Western economies by then can still stand the strain, food will be donated to help the starving. And, as happened in the Sahel, that will further depress the economic demand for food, further undermine the farm economy, and worsen the basic situation.

Kenya unfortunately is not alone in its plight; many poor countries have very similar problems, sometimes worse ones. Kenya's population growth is the world's fastest at the moment, but several of its neighbours are not far behind. On the other hand, Kenya has some advantages: its tourist industry is fairly well developed and brings in foreign exchange; its population is relatively healthy and on the way to being mostly literate; roads and other communications infrastructure are well developed; and the government has been fairly stable, especially by African standards. Yet, unless the runaway population growth can soon be curbed, the economy can somehow accommodate the rising flood of job seekers, and the agricultural sector can be made more productive on a sustainable basis, Kenya is heading for disaster.

It should be obvious that a serious effort to reduce the population growth rate would pay large dividends in easing Kenya's problems. A family planning programme with government support has existed in Kenya (and most other poor nations) since the 1960s. But in Kenya and throughout much of sub-Saharan Africa, the government support until recently consisted more of lip service than whole-hearted backing. Pronatalist attitudes are very strong in African cultures; the status of both men and women is heavily tied to the number of their children. Together with lingering high infant mortality rates, these attitudes have kept the idea of family limitation from gaining wide acceptance. At the same time, traditions that tend to moderate fertility (such as prolonged breastfeeding and abstinence from sexual relations during the postnatal period) have largely been abandoned.

Most African leaders have shared these attitudes and generally tolerated family planning only as a means of improving the health of mothers and children (for instance, by spacing births). They also felt that their nations were, if anything, underpopulated and were therefore complacent about rapid growth rates. Only since the enormous tragedy of the Sahel famine drove home the lesson of overpopulation have African governments, including Kenya's, begun to think seriously about reducing birth rates. But now the momentum of population growth is so great that putting on the brakes will almost certainly take at least a century, during which many of these populations may triple or even quadruple in size (assuming that death rates can be held down).

The populations of most nations on Earth are not expanding so fast, fortunately. The success of death control measures has come slowly in Africa south of the Sahara, and this no doubt has discouraged the

51. **Breast-feeding and fertility.** Breast-feeding is the safest and most nutritious way of feeding young children, especially for mothers in poor countries where dependable supplies of clean water and baby formula are not always available or affordable. It also contributes to the health of both mother and child by preventing conception of another child too soon. Unfortunately, in many poor countries, women have been giving up breast-feeding in favour of bottle-feeding. This nurse is part of a rural health programme in Tanzania; she is advising the mother of an eight-month-old infant who shows symptoms of severe malnutrition – a wasted body, sparse hair and insatiable hunger.

acceptance of family planning there. Birth rates have fallen in many developing countries, though; in a few cases rather spectacularly (see Chapter 9). Yet the overall average growth rate of developing nations (excluding China) is still nearly 2.5 per cent a year, more than fast enough to cause many of the problems seen in Kenya.

Indeed, poverty, hunger and unemployment are not the only possible consequences of overpopulation. As a population of any species of animal increases in numbers, more and more individuals are thrust into submarginal habitats. Nowhere has this been more dramatically illustrated by a human population than in vastly overcrowded Bangladesh. With almost 2,000 people per square mile, there is a desperate shortage of land to cultivate. As a result, poor peasants are crowded into the lowest parts of the Brahmaputra-Ganges delta. It has been estimated that about 15 per cent of Bangladesh's 100 million people now live less than ten feet above sea level.

New land is appearing rapidly in the delta as a result of the transport and deposition of silt from massive erosion on the deforested slopes of the Himalayas – ironically a consequence of overpopulation in Nepal, where other poor peasants have denuded the mountains in a desperate search for firewood. As 'chars' – shifting silt bars – form in the delta

waters, Bangladeshis occupy them and begin farming. The soil is
nutrient-rich, but the land is not suitable for human habitation. It is
too low and too exposed to the notorious tidal surges that frequently
accompany violent cyclones in the Bay of Bengal and to the giant
floods that pour off the now nearly naked flanks of the world's highest
mountains.

The price exacted for living on the fringes of an overpopulated
society has been high. In 1970 a tropical cyclone (hurricane) caused a
giant tidal wave to sweep across the delta and drown perhaps 300,000
people. That disaster was repeated in 1984 and again killed tens of
thousands. In both cases, the exact casualty numbers will never be
known.

In spite of these and lesser disasters, the peasants continue to take
the risk. Within weeks of the 1970 tragedy, population pressure
pushed people back onto the very land from which relatives and
friends had been swept away to oblivion. In the delta, poor peasants
have the chance to acquire land of a quality not easily obtained
elsewhere. Even when the possibility of moving is available, it is often
passed up. Many of the Bangladeshis are long-term residents of the
area. It has become 'home'; they are held in peril by tradition and
family ties.

Bangladesh offers an extreme example of a planet-wide
phenomenon: people who must have a place to live and make a living
are forced into hazardous choices by overpopulation and by economic
systems that are insensitive both to human needs and environmental

imperatives. Consider what is happening in earthquake-prone and poverty-stricken nations in Central and South America and Asia. Overpopulation, poverty, shortages of wood and high costs of concrete and steel are leading to larger and larger slum populations housed in stone huts and loosely constructed shanties, often on steep hillsides.

The results are both predictable and tragic. In Guatemala in 1976, for example, an earthquake killed 22,000 people and injured an additional 75,000. One million of that small nation's six million people were rendered homeless. A vulnerable area in Mexico City was levelled by a quake in 1985, killing over 20,000 people and leaving hundreds of thousands homeless. On both occasions, as always, the suffering was largely restricted to the poor. As the world becomes more crowded and dangerous, the rich continue to manoeuvre themselves into positions of relative security. Their behaviour is as predictable as the inexorable collision of Earth's tectonic plates, which is the source of the earthquakes.

But the security of the rich is illusory. The many deaths and injuries suffered as a direct result of poor people being forced into marginal habitats by population pressures are tragic. But they are only the tiny

53. **Twilight at noon in São Paulo.** One of the world's smoggiest cities, as well as one of the biggest, is São Paulo, Brazil's leading industrial centre. Shanty towns can be found around this seemingly prosperous and thoroughly developed city, too, but most of them are outside the city centre.

tip of a gigantic iceberg. Humanity's expanding numbers not only have grim consequences for the least fortunate members of society; they are ultimately a threat to everyone, even to the wealthiest individuals in the stablest societies.

In poor nations everywhere, rapid population growth and lack of rural employment have led to an incredible urban explosion, like that in Nairobi, as millions of peasants are squeezed off the land. Population growth rates in third world cities are often twice that of the nation – from 6 to 10 per cent a year. A few decades ago, nearly all the world's very large cities (over 5 million residents) were in developed countries. By the end of this century, the majority will be in developing nations, and their populations will be much larger than those of cities in developed regions have ever been.

Mexico City, already the world's largest city and all but unmanageable, had 18 million inhabitants in 1986 and is projected to reach 26 million by the year 2000. Attempts by the Mexican Government to stem the tide of urban immigrants through decentralised planning for industry have had little perceptible effect so far. São Paulo, Brazil, whose ranks of high-rise office and apartment buildings sprawling across a valley several dozen miles wide, already seem even to dwarf Manhattan's, is not far behind. In both cities, the imposing buildings, representing a prosperous developed sector of the economy, are interspersed with appalling slums and evidence of widespread poverty.

54. **Community television.**
Modern communications such as this communal television set in Port au Prince, Haiti, bring information
about life in rich countries to people in developing nations.

The very fabric of the world's social structure is threatened as overpopulation, urban migration, and (to some degree) automation interact to swell the ranks of the unemployed. The position of the rich is increasingly precarious everywhere, as the problems of the poor have spilled over into the industrialised world. One obvious – and extreme – result of frustrated expectations, of course, is terrorism. Political unrest and frequent revolutions are another, although at least these represent an attempt to improve things in the home country. A third response is a direct if relatively unnoticed invasion of the sanctuaries of the rich by the migrating poor.

55. **Riots and revolution.** Poverty and hopelessness have many times sparked revolutions. This riot occurred in Port au Prince, Haiti, shortly before Baby Doc Duvalier was deposed as president in 1986. Haiti, the poorest nation in the western hemisphere, was looted and exploited for personal gain by the oppressive Duvalier regime until the people rebelled. Many desperately poor Haitians also emigrated, often illegally, to the United States.

Displaced peasants and unemployed urban workers in poor countries often move on in search of better opportunities, which are to be found mostly in more developed regions or countries. Thus southern Africans have migrated to work in South African mines and factories; west Africans have gone to more prosperous Ghana and oil-rich Nigeria; Colombians move to Venezuela; Central Americans go to Mexico; they and Mexicans both migrate to the United States (legally or illegally); southern Europeans and North Africans have moved to northern Europe, as have sub-Saharan Africans, East and West Indians, Pakistanis and Middle Easterners. Jobs have been comparatively abundant in the richer countries, particularly for poor, relatively undereducated migrants who are willing to take low-paying jobs.

Migrants from less developed areas to Europe were encouraged in the expansive 1960s and early 1970s. When harder times arrived in the mid-1970s, partly triggered by the rise in energy prices, foreign workers were no longer welcomed; on the contrary, many were sent home – to add to the burdens of their home countries' already faltering economies. Immigrants similarly have been returned to their native homes, sometimes forcibly, from many developing nations, including Honduras, Ghana, Nigeria and South Africa.

56. Illegal immigration. The disparity between rich and poor countries is a major impetus for inter-national migration, which is sometimes done illegally. These undocumented migrants attempted to cross the border from Mexico to the United States during the night and were arrested by the US Border Patrol. In recent years, a million or more illegal immigrants have been arrested in the United States and returned to their home countries each year. The 'problem' of illegal immigration has been created in part by the policies of the United States towards its southern neighbours over the past 150 years.

The traditionally heavy immigrant stream to the United States has continued, swelled by refugees from the political strife in Central America and South-East Asia and with a large component of illegal border-crossing. Unlike European nations, the United States shares a long border with a developing nation – Mexico. While the Canadian border remains unfortified, the Mexican border is guarded and patrolled – and leaky. It is crossed by hundreds of thousands of undocumented immigrants each year, most of them simply seeking jobs.

The employment situation in Mexico is especially grim. An estimated 50 per cent of the labour force is unemployed or underemployed when times are good. As in Kenya, the labour force in Mexico has been growing even faster than the population as a whole because of recent high fertility and the resultant heavy preponderance of young people. In 1985, 42 per cent of the population were under 15 years of age (lower, however, than ten years earlier, when the percentage was 49). By the mid-1980s, times were anything but good for Mexico. Mounting massive debts (nearly 100 billion dollars in 1986) and the loss of income from oil production thanks to the worldwide glut and price drop, were pushing the country rapidly toward bankruptcy. The result, understandably, was a surge in migration northward.

Unemployment problems are not limited to the poor nations, of course. The United Kingdom's unemployment rate has hung stubbornly around 13 per cent for several years. Many young people go straight from school to unemployment benefits, where they have little hope of promotion. The discipline of regular job-holding is an unknown aspect of their culture. Street crime, drugs, and political and social unrest are the predictable consequences, and the 1985 soccer riots an obvious symptom. What will happen to the current generation of never-employed when the proceeds from North Sea oil can no longer support generous unemployment benefits is anyone's guess.

57. **Hard times in London.** All is not rosy in rich countries. These homeless young people were living in the streets in London in 1983. They had no jobs and little prospect of employment.

Things are not much better in the United States. The unemployment rate has ranged between 7 and 11 per cent in recent years, but reportedly these figures do not include the 'hard-core' unemployed who have simply given up looking for jobs. Even though low-paying menial jobs apparently exist for illegal immigrants (who are less demanding of fringe benefits and job security guarantees), they are scarce for some other categories of workers.

The ranks of the unemployed have been joined in this decade by a new underclass: the homeless. These people (a few of whom even hold respectable jobs) have through various misfortunes lost their homes and cannot afford to re-enter the housing market or even to rent suitable housing. In Britain, abandoned housing has been taken over

58. **Hard times in Chicago.** A soup kitchen at a rescue mission provided meals for homeless street people in 1985. The need for such charitable services has dramatically increased in the United States since 1980.

by homeless squatters; but too little of that remains in the United States, thanks to the American penchant for demolishing 'unused' buildings. So the homeless in the United States can be seen living on the streets in every large city, gleaning their food from dustbins, sleeping on park benches or in bus stops, and huddling in winter on warm air exhaust vents outside buildings.

West Germany's unemployment problems have been comparatively minor, but nevertheless have been sufficient to induce the government to send some of its 'guest workers' back to their home countries during recessions, especially in 1975. Since then, few if any more foreign workers have been admitted to the country. Switzerland, too, has gone through periodic spasms of trying to eject its guest workers through legislation; but each time it has been pointed out that certain sectors of the economy are dependent on them.

Today, only Japan, Hong Kong, Korea, Singapore and a few other bastions of high technology and cheap labour seem immune to problems of unemployment. But that situation is clearly fated to change. Japan, for example, is resource-poor and extremely vulnerable to disruptions in the inflow of raw materials on which the Japanese are so dependent, as the oil crises of the mid-1970s demonstrated. Such disruptions seem sure to occur more and more frequently in the future, as Earth's riches are further depleted and dispersed, and as competition for the remaining deposits intensifies. Maintaining steady supplies of essential resources seems bound to become increasingly difficult, especially for nations like Japan that do not possess them within their own boundaries.

59. **Intensive land use in Japan.** The population of Japan, some 120 million, is squeezed into a mountainous chain of islands whose area is smaller than the size of California. With few resources and little land level enough to farm, Japan's use of cultivable land is extremely intensive, as this aerial view taken near Tokyo in 1973 shows. Sunlight is reflected from the water in rice paddies.

Japan is also extremely vulnerable to the consequences of economic revenge in the form of trade barriers erected by its trading partners. Many nations have felt the lash of Japan's trade policies, which combine a coordinated government-industry economic imperialism with a subtle protectionism. Either a return to protectionism worldwide or a new energy crisis, both quite likely in the next decade, could leave Japan a poverty-stricken, hungry shambles. With this prospect, the fears of some analysts that Japan might rise once more as a military power, allied this time with the People's Republic of China, seem not too far-fetched.

There are other threats to the security of the rich posed by overpopulation besides economic and political instability. One is the increased vulnerability of the entire vast population of *Homo sapiens* to epidemic disease. Large numbers of poorly nourished people living in crowded, unsanitary conditions with inadequate medical care are a prime breeding ground for the spread of infectious diseases. This vulnerability applies whether the disease is transmitted directly between individuals (as are influenza and measles), through food or water (cholera, typhoid, hepatitis), or through an intermediate host (malaria, yellow fever, plague and schistosomiasis). The possibility always remains for the resurgence of long familiar, though mostly controlled diseases such as yellow fever and polio, or of frequently mutating ones such as flu (which after all killed some 18 million people in 1918–19).

Malaria, ancient scourge of the tropics, was largely suppressed for decades by the widespread use of insecticides to control its mosquito carriers and by the use of powerful anti-malarial drugs as both preventatives and cures. Today it is enjoying a comeback. Mosquitoes have become resistant to one after another of the series of pesticides deployed against them, while the plasmodia (tiny protozoans, relatives of amoebas) that cause malaria have similarly developed resistance to the anti-malarial drugs. Rising death rates due to malaria – not to mention a higher incidence of debilitating illness – have occurred in numerous tropical regions, including India, South East Asia, Brazil and much of Africa. The disease has also reappeared in areas that were free of it for decades.

60. **Malaria-carrying mosquito.** A mosquito of the genus *Anopheles*, which can transmit the widespread disease malaria, bites a human host. The red blood from the host is visible in the mosquito's abdomen. A recent resurgence of malaria – which affects more people in terms of debilitation and death than any other disease – may have produced nearly 300 million cases in 1985.

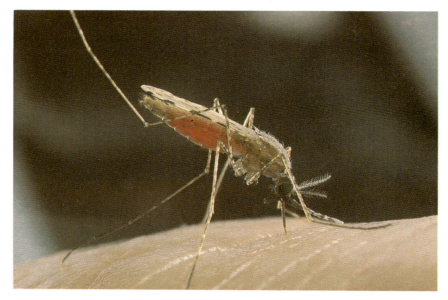

It remains to be seen whether a vaccine can be developed and deployed before humanity loses its chemical battle against malaria. New anti-malarial drugs are continually being introduced, but the plasmodia quickly evolve resistance to each one. And there may be a trend toward greater toxicity to people in the drugs. Unlike bacteria, plasmodia are eukaryotes, whose physiology is quite similar to that of other eukaryotes, including people. To cure malaria, we basically must begin poisoning ourselves, choosing chemicals to which the plasmodia are more susceptible, and hope to kill them before we kill ourselves. One recently introduced, widely used anti-malarial drug kills about one out of every 20,000 people who take *just the preventative dose*.

There is also the possibility of a completely new disease arising in the human population and developing into a major epidemic before medical defences can be mobilised. Indeed, such new diseases have appeared several times in recent years. Two of these occurrences were no more than close calls, and the public is largely unaware of them. But the third is already causing widespread alarm where numerous cases have developed – and with good reason.

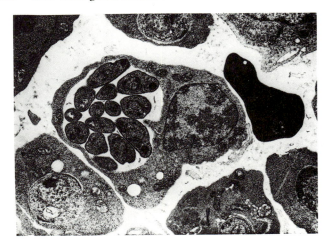

61. **Plasmodia in blood cells.** The malaria-causing organisms are protozoa (single-celled eukaryotes) known as plasmodia, and are shown here infecting a human red blood cell in a photo taken with an electron microscope. Four different species of plasmodia infect human beings; all cause debilitating disease and one (*Plasmodium falciparum*) is often fatal. Malaria kills an estimated one million children under the age of five every year in Africa alone.

The first of the two minor cases was a severe haemorrhagic disease called Marburg virus, named after the German town where it appeared and was, fortunately, contained. The virus was caught in 1967 by 25 laboratory workers from vervet monkeys shipped in from Africa for use in medical research. Five other people contracted the disease from human patients. Seven of the first 25 victims died, even though all were in a position to receive excellent medical care. Only a short time before the outbreak in Marburg, the monkeys had passed through London airport, where potentially they could have infected hundreds of travellers before the disease could even be identified, let alone diagnosed and treated, possibly launching a global epidemic. That it did not happen was a matter of sheer good luck.

The second event was the appearance in remote western Africa of a strange new disease called Lassa fever – also named after the village where it was first discovered in the late 1960s. Lassa fever, also a virus, and also fatal to a large proportion of its victims, seemed to come from nowhere. It was eventually learned that the disease was endemic among small rodents – rats and mice – and the people had picked it up from them. From Africa, Lassa fever was spread through evacuated missionary medical workers to the United States. But the world's luck held again. As it passed from one human victim to another, the virus progressively lost its virulence. Then a serum was developed from antibodies produced by survivors, so even if the disease recurs in the human population, it can be contained.

Such luck cannot always be counted on to save us. In the late 1970s, another new disease appeared – acquired immune deficiency syndrome (AIDS). This one is also a virus, one with special properties that make it unusually hard to combat. The AIDS virus attacks the immune system, making victims with the active disease susceptible to a variety of tumours and infections which ultimately cause death. An infected individual may carry the virus without developing symptoms for many years, possibly a full lifetime. Yet, even though free of symptoms (and quite likely unaware of being a carrier), an infected person may be able to transmit the virus to others.

62. **AIDS** – the newest disease to threaten humanity. An early victim of the AIDS epidemic in the United States was film and television actor Rock Hudson, shown here in photos taken before he contracted the disease and shortly before he died of it in 1985.

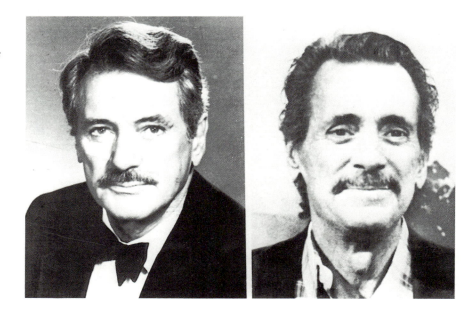

But the AIDS virus is not transmitted through casual contact as are most familiar viruses such as colds, flu, or measles. It seems to be transferred between people through body fluids, especially semen and blood or blood products. In this respect it resembles hepatitis B, which has commonly been spread among drug users, haemophiliacs and recipients of transfusions via contaminated needles. AIDS is found among these groups as well. But the subgroup of Western populations that has been most at risk is male homosexuals, apparently because of their sexual practices. Transmission between men and women through sexual contact is also possible and becoming more frequent, and infants can acquire the virus from their mothers before or during childbirth.

Among homosexual men in the United States, occurrence of the disease had reached epidemic proportions by the mid-1980s in several cities; it was also rising rapidly in the United Kingdom, Australia and western Europe. But the disease is widespread in central Africa, where it seems to have originated. It is suspected that the AIDS virus is a relatively benign disease of green monkeys that has somehow been transferred to the human population — where it is anything but benign.

In Africa, though, AIDS is not primarily confined to male homosexuals but occurs with roughly equal frequency in both men and women. It still seems to be transmitted mainly through sexual contact or contaminated needles. Why the disease is more readily transmitted between heterosexual partners in Africa than elsewhere has been a matter for speculation but remains unclear. One possible important factor is the prevalence in African populations of malnutrition, which is well known to suppress resistance to disease.

63. **A green monkey.** It is believed that AIDS is caused by a retrovirus which was originally transferred into the human population as a mutated form of a relatively benign disease of green monkeys in Central Africa. It is possible that increased contact between people and the monkeys as a result of extensive deforestation may also have been a factor in the transfer. This juvenile green monkey was photographed in the Abuko Nature Reserve, Gambia.

Certain characteristics of the virus have greatly hindered efforts to control it. Like most viruses, it is unaffected by antibiotics. Moreover, its genetic structure appears to change frequently, which makes it difficult to design an effective vaccine against it — a characteristic shared with flu viruses. Since the antibodies of AIDS carriers seem unable to wipe out the disease completely, an antibody serum appears unlikely to be useful. And, while it attacks primarily specialised blood cells involved in immune responses, the AIDS virus can also invade and damage the brain; indeed, this manifestation of the disease appears to be dominant in the African populations in which it has been found. But the brain is a protected area not easily penetrated by antiviral drugs. Because of these difficulties, it is uncertain whether an effective vaccine or cure will ever be developed.

Much of the public in western nations has regarded AIDS as a problem of certain subgroups of the population that they would just as soon see disappear, rather than as a potential threat to themselves. Some even seem to see it as a sort of just punishment for sins. Others, particularly members of groups at higher risk, have reacted with fear, often uninformed fear. Thus parents have induced schools to keep out children who were carriers of the virus (but not manifesting symptoms). In one extreme case, a child was excluded from school because a sister had the disease (although there was no evidence that the excluded child had even been infected).

In cases of active disease, keeping a child out of school makes sense, for his or her own protection if nothing else in view of the AIDS victim's increased susceptibility to infections. And a case might be made for excluding a carrier as well, since young children normally play roughly and often inflict cuts and scratches on each other. While casual contact seems extremely unlikely to result in passing the virus, blood to blood contact – direct contact of open cuts, sharing a sink while washing hands with scratches – possibly can.

Dealing with AIDS is thus proving to be a major social problem, in which the civil rights of the victims have become an issue. This is mainly because AIDS is so different from other, familiar virus diseases, because little is yet known about it, because the course of the disease is slow, and because, once symptoms have appeared, the disease has invariably been fatal. Whether all people who have the virus in their bodies will eventually develop the disease is not clear; nor is the maximum length of the incubation period known, although it can evidently be many years.

In addition, the modes of transmission are not fully understood, so the best ways to avoid the virus are not entirely obvious. Some progress has been made by the development of tests for antibodies, thus making possible a partially reliable screening of blood donated for transfusions and other uses, and through modification of sexual behaviour, especially in the homosexual communities. Attempts have been made to change the behaviour of prostitutes (who are often carriers or even victims) and illegal drug users, with less success.

Yet, despite progress in understanding the phenomenon of AIDS, by 1986 the incidence of the disease was still rising rapidly, and cases had been reported in most regions of the world. The doubling rate for numbers of patients with symptoms in the United States was less than a year, with extremely high rates of exposure (above 50 per cent) being seen in some homosexual communities. Some reports suggested that as many as 10 to 25 per cent of the populations of some central African countries were affected, and the disease seemed to be spreading swiftly. African governments had barely begun to acknowledge that a problem existed, however, and assessments were largely based on guesswork and isolated reports.

AIDS has the potential to become the 'black death' of the twentieth century – wiping out a substantial proportion of the all too susceptible human population before it could be stopped. A major danger is that the highly mutable AIDS virus might change its colours again to become more easily transmissible – perhaps by casual contact. With no vaccine and little medicine to use against it, the disease could sweep around the world, carried from place to place in hours by jet travellers. Malnourished populations with little resistance in poor countries might be most susceptible, but it is already clear that the rich are not immune.

At present, far too little in the way of resources is being invested in understanding the disease and the danger it poses to the entire world community. That AIDS appears to be rather commonly passed between heterosexual partners in African populations should be ringing alarm bells everywhere. The complacent beliefs by many leaders in the developed world that the disease attacks only certain kinds of sinners and unfortunates and that modern medicine has permanently triumphed against the ancient evil of disease could prove disastrously misplaced.

But beyond the relatively evident dangers of epidemics posed by the existence of an unprecedentedly large and fast-growing population, there are also more subtle social consequences, many of which are felt in rich countries as well as poor ones. Although these social effects may be difficult to evaluate, they could be just as important as the obvious ones of hunger and poverty. One such consequence is the loss of individual freedom, which inevitably accompanies increasing population density.

The most obvious loss of freedom from population growth in democratic nations is the dilution of each voter's impact. Whatever the system of election, the value of each individual vote declines as the number of voters increases. Moreover, the ability of a government to respond to the needs of each citizen generally declines as the number of citizens increases, once the population has grown large enough to exhaust the possible economies of scale in provision of government services.

More and more people want their cases heard by the courts, or want the toxic waste dump that is poisoning them cleaned up, or want assurance of access to employment, decent medical care and a secure retirement. But for many, these wants cannot be satisfied – and the trends indicate that the ranks of people unable to obtain satisfaction from their governments are growing.

64. A queue in Poland.
Beyond a certain point, efficiencies of scale are no longer generated by increased population size. Other things being equal, increasing the numbers of people beyond the minimum needed to run a modern society normally diminishes the freedom of individuals.

In the 1960s, C.P. Snow commented on the declining quality of telephone services: '... the difficulties of a service increase roughly by the square of the number of people using it.' 'Snow's Law' applies in many areas of society. Simple mathematics tells us that for N people, there can be $N(N-1)/2$ connections or relationships between them. For two people the number of connections is one, for three it is three, and for four, six. The number of connections increases very much more rapidly than the number of people: ten people may have 45 connections; 100 people 4,950 connections; 1,000 people, 499,500 connections; and a million people, potentially half a trillion connections. This relationship may well be a significant factor in many of the problems seen in the functioning of large modern societies.

Freedom from want, which increased throughout most of this century, now seems to be on the decline, even in some wealthy countries, as witness the stubborn unemployment, homeless people and returning hunger in the United States. The prospects are that freedom from want will plunge in many poor nations as the unemployment picture worsens. In Mexico, for instance, where unemployment is already about 50 per cent, jobs will have to be found in the next fifteen years for roughly 20 million more people (33

million of 80 million Mexicans in 1985 were under 15 years of age),
even if most young women do not enter the work force.

The situation in Mexico is typical in developing countries, where
40–50 per cent or more of the eligible workers are unemployed or
underemployed. With populations doubling in less than thirty years,
even if birth rates miraculously dropped immediately to replacement
level, the employment situation faced by poor countries would remain
overwhelming. Roughly one billion people *already born* in the
developing world are under 15 years of age. Even if few of the women
enter the work force (a very conservative assumption), several
hundred million new jobs will have to be created in poor countries
over the next fifteen years *just to perpetuate today's level of unemployment.*

65. **A crowded beach in Japan.** Population growth multiplies the potential connections between individual people, thereby contributing to the complexity of modern societies. Japan has long had a high population density and has developed some interesting social mechanisms for coping with the enormous numbers of people each person encounters every day. This is Kamakura Beach, south of Tokyo.

Even the more slowly growing population of the People's Republic of
China (current doubling time about 65 years) includes about 340
million people under the age of 15. Thus the rough equivalent of the
entire populations of the United States, Great Britain and Canada will
(allowing for some mortality) reach working age in China before the
year 2000. Meanwhile, the proportion of the work force dying or
retiring in old age is comparatively minuscule. Clearly, brute numbers
alone are increasingly restricting people's freedom to find employment.

Much of the impact of population growth on freedom is less direct, however. Once the population of a nation (or a province, or a city) is past a certain size, numerous restrictions on freedom are necessary for the common good. On thinly populated frontiers, sewage from a few people can be dumped into a stream without overwhelming its capacity to self-purify; traffic can move at random over the countryside; there are few or no neighbours to be disturbed by loud noises; and epidemics are not easily propagated because crowds seldom form and people come into contact with each other infrequently.

In Los Angeles, New York, Toronto, London, Hamburg, Rome, Moscow, Tokyo or Sydney, by contrast, laws are required to regulate matters of public health, safety and tranquility. Laws on littering, disposal of sewage and toxic wastes, keeping dogs on leashes, smog and noise abatement and the use of massage parlours all limit freedom, at least partly in response to the consequences of people living at high density. And, in general, the higher the density, the more necessary and restrictive the laws must be.

Of course, many freedoms are curtailed because populations have grown beyond the level of economies of scale. These include the freedom to drive on uncrowded streets and highways, to find solitude in the wilderness, to hunt or fish for sport and the freedom to plan one's own future. These freedoms have not all disappeared, but they are usually surrounded by restrictions of various sorts or can be exercised only under crowded, less than optimal conditions. And, of course, many of these freedoms are not available for most of the world's people in the first place.

Governments confronted with population-related problems often deal with them in ways that limit freedom. Many of Great Britain's horde of unemployed youths, lacking productive outlets for their talents and energies, rioted at soccer matches, leading to restrictions on the public's freedom to attend those matches. As more and more Latin Americans stream across the southern border of the United States without legal documentation, the debate intensifies over the need to issue identity cards for all American citizens. As drought and famine stalk Ethiopia, hungry, homeless refugees are herded from camp to camp or forcibly relocated in inhospitable areas.

None of these or many similar phenomena can be blamed entirely on overpopulation, but the increasing outstripping of Earth's physical, biological and social resources by *Homo sapiens* has substantially contributed to them and has certainly hindered their solution.

A dramatic example can be seen in the rise of terrorism. In itself, terrorism practised either by individuals or states is nothing new; and its roots in poverty, despair, greed, racism, religious prejudice and unfulfilled economic and political ambitions are hardly novel either. But now poverty-stricken, politically disenfranchised people exist in greater numbers than ever before. And many nations, such as Israel, South Africa and Nigeria, are experiencing severe disparities in population growth rates between religious and ethnic groups. The more rapidly growing subpopulations are often relatively poor and powerless, and are usually perceived as a threat by the groups in power. Population growth, differential population growth and dashed expectations are thus watering the roots of terrorism as they have never been watered before.

66. **Airport security.** The rise of terrorism and the intrinsic vulnerability of highly centralised modern societies has necessitated the establishment of increasingly strict security measures to protect travellers, who have had to give up some of their freedom in exchange. These armed guards are at London's Heathrow Airport.

At the same time, the ease of transport, increased crowding, centralisation of the machinery that supports society (such as the trend toward gigantic power plants) and arms sales to developing nations, all themselves at least related to population growth, are making innocent people everywhere ever more vulnerable to the acts of terrorist groups. The resulting restrictions on liberty are suffered at every airport and at the entrances to public buildings in many nations.

Loss of freedom and the dangers posed by potential worldwide epidemics and destabilised economies are seldom connected in the public mind with overpopulation. Even the prevalence of poverty, hunger and unemployment is often seen merely as a problem for poor countries that are having difficulty 'developing' their nation. Rapid population growth may be hindering their efforts, but it is believed that a much larger population can be supported when development is accomplished and agriculture is modernised. After all, the global food harvest is still increasing more rapidly than the population, and the

105

green revolution has barely begun in much of the developing world.

It is frequently asserted, moreover, that if everyone on Earth had a 'fair share' of today's food production, no one would be hungry. Indeed, if the rich did not insist on eating so much red meat and high-cholesterol dairy products (which are unhealthy for them anyway), much less grain and protein-rich oilseed presscake would have to be fed to cattle. Grain now fed to bovines could then be made available to the starving people in Africa, and land used to produce animal feed could be diverted to providing direct sustenance for human beings. By this reckoning, Earth is not overpopulated – it's just mismanaged by the selfish, ignorant or uncaring. Maldistribution is the culprit, not too many of God's children.

It sounds good if you say it fast, especially since it is true, at the moment, that Earth's yield is sufficient to feed everyone, and, if distribution could somehow be equalised, no one would go to bed hungry. It is, of course, equally true that manufacturing nuclear weapons carries no threat to civilisation because, if all human beings loved one another, there would be no chance of nuclear war. Both positions are preposterous because they deal not with humanity, but with an idealised *Homo superior* – a species of saint that shares equally and loves everyone equally.

In the real world, food (to say nothing of other material goods) is *never* shared equally on the basis of, say, idealised body weight, age, sex and basal metabolism. There is and always has been gross maldistribution between rich and poor nations, rich and poor areas within nations, and rich and poor people within areas. In addition, especially among the poor, there is normally maldistribution even within families. Typically, a working father is proportionately better fed than his young child. If the child perishes, that is a tragedy; if the father perishes, the entire family may starve.

We cannot plan for a world populated by saints when humanity has shown precious little tendency to change its behaviour in that direction. Indeed, one could argue the exact reverse. Maldistribution of wealth is surely far greater now than ever before in human history. Until this century, the rich were so few in number that evenly dividing their food or other assets among the poor would have not made a discernible difference to the recipients. But today, perhaps one in four of the world's human beings enjoys an affluence undreamed of before this century, while two others just get by, and the fourth lives in the most abject poverty on the verge of starvation. By and large, for the last several decades, the rich have continued to get richer, while the poor have mostly remained poor (they could hardly get more so).

Of course, some wealth has been redistributed to the poor, as testified to by the existence of multitudinous aid programmes,

67. **Ultimate urban poverty.** Wealth is inequitably distributed within nations as well as between them. In a slum in Cairo, Egypt, families live amid the squalor of the garbage tip. Often in poor countries, garbage tips support many destitute people who forage for food and other materials in them. Television viewers in rich countries were shocked in 1986 by scenes of hungry Philippinos fighting for access to fresh garbage as it was delivered to a dump in Manila.

charitable organisations and activities. But much of the money invested in these programmes returns to the rich in the form of administrative payments and other benefits, such as required shipment of food and other aid on the carriers of donor nations. The benefits that actually flow to the poor are a trickle compared to the actual need, especially compared to the wealth of the donors. And the assistance and donations sometimes do more harm than good, as when donated food depresses the agricultural economies of poor nations.

The rich countries, moreover, are unlikely to increase their assistance substantially. The purse-strings tighten whenever economic conditions deteriorate, even though the poor suffer most in hard times. But conditions are unlikely to return to the heady global expansionism of the 1950s and 1960s. In addition, the more terrorism rears its ugly head – an all-too-human response to the gross inequalities of modern life – the less the rich feel like giving and the stronger the urge to protect their interests and property.

Above all, even if by some miraculous stroke of enlightenment, humanity were able to manage a totally equitable worldwide distribution of Earth's yield, the solution would be temporary. That yield is now being maintained only by the profligate use of non-renewable, finite resources such as arable soils and aquifers. Meanwhile the human population continues to expand.

It seems reasonable, given these realities, to adopt a biological definition of overpopulation – one that deals with the organisms as they are, not as they hypothetically might be. If the population of lions on the Serengeti Plain grew so large that most of the antelopes were eaten and the lions began to starve, biologists would judge them to be overpopulated – even though more lions might be supported if they evolved overnight the desire and ability to eat grass. Similarly, if there are so many people in the Sahel that local resources cannot support them all *with* the maldistributions that characterise all human societies, then that region is clearly overpopulated.

The basic point is that overpopulation is not just an issue of how many bodies there are relative to resources and values. It also has to do with how people organise their societies and what they want out of life. Overpopulation could theoretically be reduced in the Sahel with no change in human numbers, by changes in political arrangements and patterns of food distribution, combined perhaps with certain changes in values. Herds, for example, might have to be reduced and grazing patterns altered to reduce the pressure and permit recovery of rangelands. Some kinds of ecologically sound agricultural development, carefully introduced, could also help. Yet a gigantic gap exists between what can be done in theory and in practice.

One of the few cheering things to come out the great Sahel tragedy has been the compassionate response of rich western countries in sending aid. But compassion is not enough. The millions of dollars worth of free food poured into starving African nations, through the efforts of empathic rock musicians and others, certainly saved lives – at least temporarily. But think of the impact of that free food on an already staggering farm economy! As is so often the case, the compassionate approach carries with it the potential for exacerbating disaster.

Should Westerners have hardened their hearts and let those wide-eyed children starve? We hope that for most people such a choice would be unthinkable. And fortunately, it is not a choice we have to face. Instead of uncoordinated food aid alone, the rich nations should be cooperating with African countries simultaneously to feed the starving, to encourage and improve indigenous agricultural production and to bring a halt to population growth (and an eventual *reduction* in population size) by humane means.

Such action would require dramatic shifts in the thinking of world leaders. The Soviets would have to cease *extracting* grain from Ethiopia in payment for arms shipped to its Marxist government. The United States would have to forget its distaste for Marxism and supply appropriate aid to people in poor countries regardless of the form of government they happen to live under. The Pope would have to inform himself about population problems and stop preaching nonsense about birth control. As will become clear, such an approach, seemingly so idealistic, is actually hard-nosed realism.

CHAPTER FIVE

Fouling the Nest

The great discrepancy in demographic situations between rich and poor nations has led some people to conclude that overpopulation is largely a problem of the misguided citizens of the Third World. This notion will not stand up to even relatively casual examination. As we shall see, *the entire planet is overpopulated*. Virtually everywhere, numbers of people are pressing on both values and the capacity of life-support systems. The human family as a whole is behaving in a way no sane individual human family would – it is simultaneously using up its capital and destroying its sources of income.

And the rich are showing the way. This is easily seen by recognising that the level of overpopulation is not determined by numbers of people alone, but by how those people behave. A lot more people can be supported in a given nation if families live at a subsistence level as vegetarians than if each family 'requires' a large air-conditioned, appliance-loaded house, several cars, access to transportation in jet aircraft and a diet rich in animal protein.

One can think of a human population's impact on its life-support systems as being roughly the *product* of three factors: the number of people, the per capita level of consumption (affluence) and some measure of the average amount of environmental damage done per unit of consumption. The first two factors are rather intuitively obvious – a small number of people living extravagantly can use resources and degrade environmental systems as much as a relatively large number living frugally.

The third factor – the *technology* factor – simply recognises different ways of living high or being frugal. If affluence is enhanced by consuming fine wines which grow on rocky soils not much good for other crops, the third factor will be smaller than if affluence means buying and driving large cars. If frugality means inefficient subsistence farming which degrades the soil, the impact per unit of affluence will be greater than if frugality means intensive organic farming. The third factor basically measures how environmentally benign or destructive are the means people use to achieve their standard of living.

109

68. An oil field with flares. Energy use is often closely connected to a society's environmental impact. In the Habshan oil field in Abu Dhabi, shown here, the natural gas that occurs with the oil is being flared off and wasted, because the cost of retaining and using it is 'too high' compared to its value in the energy-rich Middle East. Burning the gas also produces a good deal of air pollution, as can be seen.

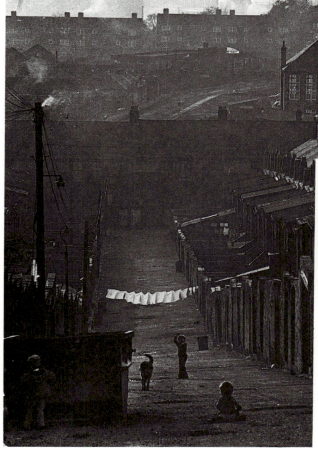

69. Industrial blight.
Visible air pollution and general deterioration of the human environment are evident in this photograph of Newcastle-upon-Tyne taken in 1976. The environmental costs of industrialisation are no more evenly distributed than are the benefits, and the highest costs are often paid by people who benefit less. The rich can afford to live upwind and away from the smoke; the poor frequently must live in it.

70. Climax, Colorado. Mining operations can cause enormous environmental damage, as this aerial view of a molybdenum mine at Climax, Colorado, indicates. Molybdenum mining is particularly destructive because the metal occurs at concentrations of less than 1 per cent, and vast amounts of rock must be ground up in order to extract it. This mine has been operating since before World War 1.

Once this simple multiplicative relationship between population size, affluence and technology is understood, then the disproportionate contribution to overpopulation made by citizens of rich countries becomes evident. For convenience, we will ignore the often difficult-to-measure third factor for the moment and focus attention on the first two: population size and affluence.

A reasonable approximate measure of affluence is per capita gross national product (GNP). In 1986, the per capita GNP of the United States was about $14,000, that of the United Kingdom about $9,000 and that of India about $260. Assuming that the impact per unit of consumption was the same in all three nations, the birth of an American baby would represent more than 50 times, and that of a British baby nearly 35 times, as great a threat to global resources and environmental systems as that of an Indian baby. A similar calculation makes a Swiss baby more than 125 times as large a contributor to the degradation of Earth's resources and environment as one born in Bangladesh.

If the difference in population sizes is considered, the nation of Switzerland still has a total impact eight times that of Bangladesh, even though Bangladesh has 16 times as many people. Similarly, India's total impact is less than twice that of Switzerland, although there are more than 100 times as many Indians as Swiss. And Japan has four times the impact of China, even though its population is only about one-ninth as large.

By this simplified calculation, the United States is the world champion, having 17 times the impact of India (with a third as many people) and 11 times the impact of China (with less than a quarter of the population). The impact of the United States is even twice that of the Soviet Union, its nearest national competitor in wrecking the planet, in spite of Russia's slightly larger population and almost equally massive industrial plant. Even the aggregate of the European Community (in 1985 consisting of Belgium, Denmark, France, the Federal Republic of Germany, Greece, Ireland, Italy, Luxembourg, Netherlands, Portugal, Spain and the United Kingdom) has only about four-fifths the impact of the United States, although it has about 35 per cent more people.

So it is the rich who disproportionately threaten the capacity of Earth to support humanity and all the other organisms that live on the only known habitable planet. It is they who are extracting and dispersing non-renewable resources and consuming fossil fuels as if there were no tomorrow while spewing pollutants into the atmosphere, rivers and oceans. They are the principal spenders of humanity's capital.

But that does not mean that the behaviour of the poor is exemplary.

While the poor may be a relatively minor threat to planetary systems (the principal exception being the contribution of poor people to the destruction of tropical forests), they are a major threat to themselves, as we saw in the last chapter, and to the stability of international economic and political systems.

71. An automobile graveyard. These wrecked vehicles about to be pulverised are symbolic of the often wasteful resource consumption of people in rich countries.

Here we will examine some of the ways in which the activities of large numbers of human beings degrade the environment, and the ways in which that degradation *directly* injures human beings, by assaulting them directly or damaging their artefacts or their crops. The next chapter focusses on the consequences of human disruption of ecosystems and how that *indirectly* poses an even more potent threat to the future of civilisation.

The most obvious and most publicised of the assaults *Homo sapiens* mounts on its environment are ones that add poisons to it – especially the spewing of the products of combustion into the atmosphere. You will recall that organisms, including people, mobilise energy to drive their life processes by a slow, controlled combustion of carbohydrates (sugars). But people are able to make use of enormously greater amounts of energy by burning various organic (carbon-containing) fuels outside their bodies.

The average person on Earth uses about ten times as much non-food energy as food energy; the average American uses about 100 times as much. And of that non-food energy, about 95 per cent is obtained by burning something: wood, coal, natural gas or petroleum products. Only the various forms of nuclear power make significant contributions to the human energy economy without combustion. Solar energy is the product of nuclear fusion (in the sun), and hydropower is a form of solar energy, since the sun's heat evaporates and lifts the water so that it can fall on uplands and run through turbines on its way back to the sea.

When food is burned inside the human body, or organic fuels are burned outside it, carbon dioxide and heat are inevitably produced. Both carbon dioxide and heat are important additions to the environment, but neither has been produced in quantities that are directly injurious to human beings. The burning of fuels, however, also releases into the environment an array of other combustion products, such as carbon monoxide, nitrogen and sulphur oxides, ethers, benzo(a)pyrene and many other chemicals. These materials are released into the atmosphere by motor vehicles, factories and fuel-burning power plants. They are also released when fuels are burned to heat homes, offices and public buildings.

These compounds can have a direct and deleterious effect on human health, as well as injuring other living things (plants, including crops, are often damaged by smog) and damaging artefacts that people hold dear. Components of air pollution can cause or contribute to diseases such as lung cancer and heart disease. While it is very difficult to estimate the cost in lives or money of air pollution, in the United States alone it is estimated to hasten thousands of people to their graves and add tens of billions of dollars annually to the national health bill.

72. **The Acropolis in smog.** Statues and stone bas-reliefs on the Acropolis, which is surrounded by the city of Athens, Greece, have been eaten away by airborne acids and are being replaced with synthetic copies. Buildings and statuary have suffered similar damage in major cities around the world.

In most developed nations in the last two decades, emissions of some of the more common combustion products from vehicles, industrial facilities and electrical power plants have been significantly reduced by improved emission control technology and the enforcement of increasingly stringent laws against pollution. But the tightening controls have to a large extent been counterbalanced by the increase in emissions from ever more vehicles, factories, offices, homes, and so forth. For instance, during the 1970s, air quality in most United States cities improved considerably, but by the 1980s much of that gain was being eroded away as the number of emission sources continued to rise, and air quality was again declining.

In all probability, the most serious near-term environmental consequence of our species' mobilisation of energy by combustion is the generation of acid precipitation. Oxides of nitrogen (produced primarily by vehicle exhausts) and sulphur (mostly from coal and oil-burning power plants, smelters etc.) undergo chemical reactions in the atmosphere and produce nitric and sulphuric acid. These powerful acids have made rain, snow and fog over wide areas of the Northern Hemisphere increasingly acidic – often hundreds of times more acid than precipitation from relatively unpolluted skies. Sometimes the acids are not formed in the atmosphere, but rather after dry acid-forming substances have been deposited on the surfaces of leaves or soil. But the results are the same: organisms in many areas are being subjected to rapidly escalating levels of acidity in their environments.

Because the chemical reactions of life require rather strictly controlled levels of acids and bases (pH, in the jargon of scientists), there are limits to the pH of the environment within which organisms can survive and flourish. Those limits vary widely from species to species, as any tropical fish fancier can testify. Conditions in which neon tetras can be bred, for example, must be much more acid than those suitable for mollies.

Some species could also adjust through evolution to large changes in pH, but that would require selection over many generations and, most likely, high mortality in each generation. But acid precipitation has been changing habitats on a time scale of a few decades, much too fast for organisms with spans of more than a year between generations (generation times) to have a chance of adjusting to the higher acidity. Moreover, the change in many areas has been much too great in scale to permit adaptation even by organisms with short generation times, such as many insects and algae.

One result of acid precipitation has been the transformation of many lakes in places like New England and Scandinavia from biotically rich and productive ecosystems to impoverished, sometimes virtually dead, bodies of water. For instance, all the fishes have been killed in hundreds of lakes in the Adirondacks, and almost 50,000 Canadian lakes are threatened with a similar fate. Brook trout are disappearing over wide areas, and the acids are adding to the burden of dams, other pollutants, poaching and overfishing that is destroying populations

115

of Atlantic salmon in Nova Scotia. In many Scandinavian lakes, populations of both fishes and the plankton species at the lower levels of the food chains that support the fishes are depressed. In desperation, governments have resorted to dumping lime into the lakes to reduce their acidity – a stopgap measure, to say the least.

Serious as the effect of acid precipitation is on aquatic systems, it may be dwarfed by the catastrophic impact on forests and other terrestrial ecosystems, which is only beginning to be appreciated. It is often difficult to judge what is happening to populations of trees, since they are organisms with very long generation times – mature trees often live for hundreds or even thousands of years. What appear to be perfectly healthy populations of live oaks in the coastal hills of California are actually gradually dying out because cattle and squirrels continually destroy the oak seedlings.

The distress of forests subjected to acid precipitation can become manifest in decades rather than centuries, but the symptoms are subtle and not easily perceived until they are rather advanced. Acidification assaults trees and other plants in several ways, but the details are hard to unravel because the plants are simultaneously being subjected to non-acid pollutants such as ozone as well as the acid compounds, and the effects of the different insults may not only be confused with one another but they may interact with each other to produce still other effects.

73. **Forest damage from acid rain.** The slow death of this forest near Ostravice, Czechoslovakia, is attributed in large part to the effects of acid rain. Such damage can be seen in wide areas of Central and Eastern Europe, and its beginnings have been detected in North America.

Acid damages the leaves themselves, reducing their photosynthetic capability and causing a loss of nutrients. Bacterial populations in the soil are suppressed, slowing rates of decomposition and the release of nutrients for the trees. Higher acidity speeds the leaching of nutrients such as potassium, calcium and magnesium from the soil, and, perhaps most importantly, mobilises toxic metals such as aluminium which are normally combined harmlessly with other soil elements. These changes in the soil are thought by some scientists to interfere with the ability of the trees to take up nutrients. Weakened by the toxic materials and the lack of essential nutrients, the trees become vulnerable to attack from insects and plant diseases.

Whatever the precise mechanisms of the damage, there is no question that increasing numbers of trees in forests over large portions of central and western Europe and in some high altitude areas of New England and the Great Smoky Mountains of the southeastern United States are dead or dying – and that the cause is almost certainly air pollution in general, with acid precipitation playing a significant role.

The syndrome, called 'forest death' by the Germans, seems to be most advanced in West Germany, where damage first appeared in the early 1970s. After 1980, the process accelerated alarmingly, as the number of forest plots that contained damaged trees expanded sevenfold between 1982 and 1984. Eleven species of trees, both conifers and broad-leaved, were affected. By 1985, trees in over half of West Germany's forest plots, covering an area of 4 million hectares, showed damage. Including several adjacent nations, trees in 6 million hectares of Europe's forest land were exhibiting symptoms, and the syndrome was spreading rapidly.

The potential for damaging or killing forests over vast areas of the Northern Hemisphere clearly exists. Symptoms of 'forest death' have been observed in forests in Eastern Europe, the USSR, Italy, Spain, Canada, Britain and the upper Midwest of America. We may now be witnessing the beginning of a widespread process, in which the most sensitive forests or those exposed to the most intense pollution are succumbing first. It is possible, for example, that acidification of some soils may become irreversible well before the trees they nourish are killed.

In a sense, humanity is running a gigantic experiment, poisoning much of one hemisphere (and probably parts of the other hemisphere), then waiting to see what will happen. No doubt, differences in the intensity of the pollution and in the natural ability of soils to resist changes in pH (some alkaline soils will acidify only very slowly if at all) will produce great geographical variation in the degree of damage and the speed at which it becomes apparent.

Acid precipitation directly injures humanity in many ways. Not only does it destroy stone carvings, it attacks the façades of buildings and other structures. For instance, in early 1986, acid rain damage to buildings in Chicago was estimated to amount to about 45 dollars per resident per year. Acidification is also reducing populations of important food and game fish, and it threatens timber and other forest-based industries, in which losses could be in the billions of pounds. Furthermore, acid precipitation may also reduce the productivity of crops. And toxic metals, especially aluminium compounds, leached from soil and water pipes by acidity, enter people's drinking water.

Steps to abate air pollution are long overdue in order to arrest the acidification of both aquatic and terrestrial ecosystems and to reduce the damage from non-acid pollutants (such as ozone) to the plants that form the basis of those systems. Not to take those steps immediately would be the height of irresponsibility, for our civilisation is dependent not only on products from the threatened ecosystems, but on the services they provide.

Human health is also threatened by several important classes of toxic substances that are not combustion products. One consists of products of the human mind – novel organic molecules produced in the chemistry laboratories of governments, universities and industries. These chemicals include a variety of chlorine-containing pesticides (chlorinated hydrocarbons) such as DDT and the herbicide 2-4D, some related compounds such as PCBs (compounds often used in electrical transformers), and two other chemical groups of insecticides, organophosphates and carbamates. Members of all three groups may produce immediate illness or death for individuals directly exposed to large doses.

But such acute toxicity is primarily of concern to workers involved in manufacturing the chemicals or using them in agriculture and pest control activities. An exception to that generality was clearly demonstrated by the 1984 disaster at Bhopal. Union Carbide of India, Limited had opened a plant to manufacture the pesticide Sevin in that city in the relatively underdeveloped state of Madhya Pradesh. The facility was welcomed, both as a source for the pesticide that was to contribute to India's green revolution and as a source of jobs in Bhopal.

But things did not go well at the plant. Drought in the late 1970s made it more difficult for farmers to purchase the expensive Sevin, while cheaper and less toxic synthetic pyrethroid pesticides began to eat into its remaining market. By 1984, the plant was operating at less than one-fifth of its capacity and cutting back its workforce.

Between 1981 and 1984, there were half a dozen accidents at the Bhopal plant, involving releases of two deadly poisons: methyl isocyanate (MIC), a compound used in the manufacture of Sevin, and phosgene, a poisonous gas used in chemical warfare. In one incident, a worker was killed, and the rash of accidents led to warnings of disaster

in the local press. But the warnings were not heeded, and on
2 December 1984 Bhopal's luck ran out. A series of mechanical and
operational failures allowed a cloud of MIC to escape from the plant,
and wind blowing towards the town and a temperature inversion
(warm air overlying cold – often trapping pollution close to the
ground) moved that cloud over several square miles of heavily
populated area. Some 2,500 people were killed, and about 50,000 were
severely injured. Many of the long-term effects remain to be clarified.

Facilities in poor countries like that at Bhopal inevitably serve as
magnets for squatters, who settle in the vicinity hoping to reap some
benefits from being near a centre of relative prosperity. Combined with
the less strict safety regulations that often characterise less developed
areas, it is a formula for disaster. In Mexico City, a liquified natural gas
facility exploded in 1984, killing 452 people and injuring ten times as
many. In the same year, a fire caused by a petrol leak in Villa Socco, a
suburb of Cubatão, Brazil, killed more than 500 people (the official
count was 93, but most of the victims' bodies were not recovered).
Cubatão is better known for being perhaps the most polluted city in
the world, with appalling rates of infant mortality, stillbirths, birth
defects, asthma, chronic respiratory diseases and skin disorders. While
disasters like those in Bhopal and Mexico City have so far been
concentrated in poor nations, the potential for them clearly also exists
in the industrialised world.

Acute toxicity, in spite of such disasters, is only the tip of the
iceberg. More worrying is the persistence in the environment of some
substances, especially the chlorinated hydrocarbons. Decomposer
organisms have no evolutionary experience with these novel
compounds and cannot easily break them down into their component
elements. Moreover, chlorinated hydrocarbons have an affinity for fat.
In the environment, they are readily taken up and retained by
organisms – including crops and animals grown for human food. The
residues of chlorinated hydrocarbons persisting in food are thus
consumed by people, and once in the body, they tend to remain there.

But the most serious effect of the gradual toxification of our planet
by human activities is not the direct deterioration of human health
that results, but the reduction of the health of the ecological systems
upon which *Homo sapiens* depends. If, for example, exposure to DDT or
other chlorinated hydrocarbons subtracts a day from the life
expectancy of the average human being (it might subtract none at all
or several weeks; scientists do not yet have the data to tell), that
would be unfortunate, but not calamitous.

Chlorinated hydrocarbons and many other toxins do have the
potential for causing calamities, however. They undergo a process
called bioconcentration, by which they accumulate in food chains and

119

build to very high levels in some organisms. Bioconcentration could lead to an ecological calamity whose consequences would dwarf those of widespread low-level poisoning of the human population.

Naïveté about the behaviour of toxic substances in ecological systems has contributed substantially to a lack of appreciation of this basic fact of biology. This naïveté is not confined to lay people. British Nobel Laureate in chemistry Sir Robert Robinson made a statement rarely matched for pomposity and ignorance when he wrote to the London *Times* on 4 February 1971: 'Neither our "Prophets of Doom" nor the legislators who are so easily frightened by them are particularly fond of arithmetic ...' He then did what he described as 'simple arithmetic' purporting to show that the amount of lead going into the oceans would be so diluted as to become biologically unimportant.

Sir Robert's arithmetic was simple all right, since it ignored the mechanisms by which living systems can reconcentrate materials that physical processes have dispersed. Thus plankton (algae and other aquatic organisms so tiny that they move at the mercy of water currents) and fishes can have 10,000 times the concentration of lead as the ocean water in which they swim. Similarly, oysters, which feed by filtering their environments, have been found with 70,000 times the concentration of chlorinated hydrocarbons in their flesh as existed in the water around them.

A major mechanism of bioconcentration is created by the structure of food chains. While the amount of biomass (living weight) declines at each transfer upwards from one trophic (feeding) level to the next, there is often no equivalent decline in the amount of toxin being passed up the food chain. Rather, compounds such as DDT tend to remain differentially in the fat-containing living parts of the system and become more concentrated as they move upwards from one trophic level to the next — from plants to herbivores, to carnivores and so on. Furthermore, predators that live high on food chains tend to live longer than those lower down — and thus have more time to accumulate poisons. The short-lived microscopic plankton in a Long Island (New York) estuary were found to contain only 0.04 parts per million (ppm) of DDT, while birds at the top of the food chains usually had thousands of times more.

High concentrations of chlorinated hydrocarbons have been shown experimentally to cause reproductive failure in birds, kill fish and inhibit photosynthesis in marine algae. Smaller but significant effects on behaviour and reproduction doubtless occur at lower concentrations. The first two effects have been seen in nature. The most dramatic consequences of the use of chlorinated hydrocarbons were observed in the 1950s and 1960s — the heyday of the misuse of DDT. Declines in populations of eagles, hawks and pelicans attracted much public attention, as did spectacular fish kills in the lower Mississippi and Rhine rivers. Human beings lost both aesthetic and food resources, but the direct effects of these poisoning episodes on people were minimal.

The human population is carrying a load of these poisons, but the significance of that load for public health is unclear. The statistical problems of documenting its health effects are daunting, since (among other things) they can be confounded with the consequences of breathing polluted air, smoking, alcohol consumption and all the other abuses to which we subject ourselves.

74. Villa Socco after the fire. This slum in Cubatão, a heavily industrialised town in São Paulo State, Brazil, was destroyed by a fire in February 1984 caused by a leaking petrol pipeline. The leaked fuel flowed beneath the village houses, which were built on stilts over the water. The villagers, rather than report the leak, siphoned off the petrol for their own use. When it ignited, the village became an inferno in seconds. People in poor countries are especially vulnerable to such disasters because they frequently live close to dangerous industrial facilities where work is available and safety standards are usually below even the often inadequate levels characteristic of rich countries.

75. Victims of the fire in Villa Socco. These bodies, charred beyond recognition, were among several hundred people killed in the tragic fire. Worldwide, one of the common penalties of being poor is to be more exposed to pollution or to human-caused or natural disasters.

121

76. Resort near Cubatão.
Only a few miles from the
poor village of Villa Socco
and the pollution of
Cubatão is this playground
for the rich, the beach
resort of Guarujá.

Controls on the use of DDT and related compounds in the early
1970s led to a decline in their presence in the environment, and to
recovery of most affected bird populations (although peregrine falcons
almost disappeared from eastern North America and are only slowly
recovering with the help of captive breeding programmes). Unhappily,
the environmental load of DDT in the United States seems to be on the
increase again, both from illegal use and from its presence as a
contaminant in other pesticides still legally in use.

Not all toxins that threaten humanity are synthetic; some, such as
lead, mercury and cadmium are simply natural elements. They may
pose serious threats today not because they are novel, but because
Homo sapiens collects them in places and quantities that are unusual or
unknown without human intervention.

For instance, civilisation appears to have roughly doubled the natural
flow of mercury from land into the atmosphere and aquatic systems.
While inorganic mercury is somewhat toxic, organic mercury
compounds are much more so. Micro-organisms in the sediments of
rivers and lakes transform relatively harmless mercury metal into the
highly toxic organic compound methyl mercury – a substance that, like
chlorinated hydrocarbons, easily concentrates in food chains. Effluents
containing mercury from a single chemical plant at Minimata, Japan
led to a horrifying poisoning episode in the 1950s. At least 46 people
died agonising deaths, some 3,500 suffered serious symptoms, and
thousands more may eventually develop problems.

Lead presents a similar story. About 15 times as much lead is mined as is normally carried to sea in the world's streams and rivers. Lead interferes with various metabolic functions, causing such diverse damage as impairment of kidney function, anaemia, mental retardation and possibly cancer. Today the primary source of exposure to lead is from car exhausts; some of the lead component of petrol is emitted with other unburned contaminants. The United States has almost eliminated lead from its fuel, but the United Kingdom and many European nations have lagged behind in their efforts to reduce it, despite clear evidence of its damaging effects, especially on children.

77. **An eagle with its prey.** Predatory birds that feed at the tops of food chains, like this Verreaux's eagle carrying a hyrax in the Serengeti plain of East Africa, are particular vulnerable to environmental poisons such as DDT that tend to be concentrated in food chains. Since DDT and some of its chemical cousins were banned in most developed nations in the early 1970s, eagles and other predatory birds have made a comeback.

Another toxic metal that has caused serious health problems for human beings is cadmium, of which about 30 times as much is mobilised by mining as by natural weathering. Cadmium is directly toxic to human beings, who are exposed to it mainly through their food. Smokers also get small doses from cigarettes. Selenium is toxic in large doses, although it is an essential nutrient in trace amounts. Selenium mobilised from soil by agricultural chemicals became a serious problem in one area of California's agriculturally productive Central Valley in the 1980s, poisoning local livestock, wildlife in a reserve and people's wells.

All of these poisonous substances can represent direct threats to human beings. They are, of course, sources of acute poisoning — causing people to choke in smog, die with a short-circuited nervous system (as can occur shortly after exposure to powerful organophosphate pesticides) or go mad and lose control of bodily functions (some of the symptoms of mercury poisoning). But many compounds are also suspected of producing more subtle long-term threats, particularly cancers.

Not all carcinogenic (cancer-producing) hazards can be laid at the

123

doorstep of humanity itself, however. Non-human organisms, as we have seen, have evolved a diverse array of poisons that are used in both defence and attack. Many of the compounds that plants produce to ward off herbivores are highly toxic, and many plant-defensive chemicals have been shown to cause mutations in bacterial test cultures and thus are at least suspected of being carcinogenic. This has led some scientists to conclude that natural substances may be the cause of many more cancers than are synthetic chemicals. Caffeine, the various plant toxins that give herb teas their flavours and scorched fats on barbecued meat may well be sources of many human cancers.

Whether this is so still needs to be investigated, however. The human evolutionary line has been exposed to natural toxins and carcinogens for millions of years and may have evolved biochemical defences against them (as our bacterial forbears did against that once-deadly poison, oxygen), whereas our bodies may be relatively defenceless against newly-created dangerous molecules. A conservative course would be to minimise our exposure to both natural and synthetic chemicals that may be harmful, while always trying to balance potential harm against potential benefits. For example, taking either the synthetic drug chloroquine or the natural drug quinine carries with it a low probability of a serious reaction to the medication, but those risks must be weighed against the very grave threat of *falciparum* malaria.

While the incidence of acute effects of synthetic toxins on people and ecosystems is not as common as it was a few decades ago, the variety of dangerous substances being released into air and water continues to rise. Except in cases where people have come into direct contact with toxic materials (such as contamination from abandoned waste dumps, as happened in the notorious Love Canal incident near Niagara Falls, in which residents had to abandon their homes), their most serious health effects probably result from pollution of water supplies.

The amount of groundwater under the United States, for example, amounts to some 50 times the annual rainfall, and about half of the American people depend on it for their domestic water. Many parts of that vast reservoir are being slowly contaminated by a wide variety of human-made poisons, largely industrial and military wastes and pesticides.

The industrial solvent trichloroethylene (TCE) has been found in groundwater in over 3,000 times the highest concentration that has been found in surface water. TCE, which is used to degrease metal, is carcinogenic in animals and is suspected of being carcinogenic in human beings. It was found, for example, in the well water of the upstate New York home of Carole Hawkins. Also in the water was another synthetic poison, vinyl chloride, a confirmed human carcinogen. Shortly after moving into the house, Carole started suffering from miscarriages and kidney problems. Her teenage daughter developed headaches and skin rashes. After some sixteen years of constant health problems, health officials discovered the contaminants in the water.

The Hawkins case is far from isolated. Viewers of the evening television news in the United States have been treated to example after example of health problems associated with contaminated water supplies. In Vermont, more than half of 50 water systems tested contained chloroform; of Battle Creek, Michigan's 30 wells, 18 were contaminated with TCE and vinyl chloride. In the San Gabriel Valley of California, some 400,000 people recently discovered they had been drinking water laced with TCE; 39 wells supplying 13 cities had to be closed down. And between 1979 and 1985, more than 1,400 wells had to be closed in California's Central Valley because of contamination with the pesticide dibromochloropropane (DBCP).

The US Environmental Protection Agency (EPA) estimates that almost two-thirds of rural Americans are drinking water poisoned with pesticides or other dangerous compounds. Indeed, the list of toxins found in American wells reads like a catalogue of an organic chemistry laboratory. A few of the chemicals not mentioned already are toluene, acetone, methylene chloride, cyclohexane, benzene, 1,1-dichloroethylene, tetrachloroethylene, dibromochloromethane, bromoform, Lindane, Parathion and alpha-benzene hexachloride. Many of these compounds are capable of producing cancers in test animals, and a few are confirmed human carcinogens, besides, of course, being a threat to health in a variety of other ways.

To make matters worse, chlorine, which is often added to water supplies to kill dangerous bacteria, also reacts with organic compounds always present in the water to form additional chlorinated compounds that may be capable of causing cancers. The EPA has recorded some 700 dangerous chemicals in drinking water and rated the levels of pollution 'serious' in 34 of the 50 states. The situation in other rich nations is similar, although in some cases, controls may be stricter. For instance, in Europe only one-quarter the level of chloroform is permitted in drinking water as the EPA allows in the United States.

Pollution in industrial areas of many less developed countries is presumably worse and regulations to control them are much more lax. As the disasters of Mexico City, Villa Socco and Bhopal have tragically demonstrated, controls on dangerous environmental situations in those nations are often nearly non-existent.

It is very difficult to forecast the health consequences of consuming chemical cocktails instead of pure drinking water, since they must be

125

discriminated from those of air pollution, smoking, food contaminants, radiation and other assaults on health. Of course, not all poisons produce obvious symptoms, such as the acne-like rash that often accompanies exposure to high levels of chlorinated hydrocarbons. When people die of cancer, one can at best make an educated guess in most cases about the cause. Moreover, the problems of determining whether cancer rates as a whole are climbing is replete with difficulties because more people are living to the age where cancer becomes common, and because doctors are more alert to it and more likely to diagnose it early.

Comparisons of cancer rates in areas with different levels of chemical contamination are not reassuring, however. White male inhabitants of New Orleans, whose water supply was discovered in the 1970s to contain 66 synthetic chemicals, including half a dozen suspected carcinogens, have 50–100 per cent more bladder cancers than the national average. Similarly, in rice-growing areas of Arkansas where herbicides contaminated by dioxin are heavily used, the cancer mortality rate is 50 per cent higher than the national average. Dioxin is known to be a potent animal carcinogen and is suspected of being the most potent of all human carcinogens.

Signs of other health difficulties, such as rashes, headaches, miscarriages and so on are often associated with the presence of toxins in drinking water. And the fertility of American males in general seems to be dropping. In 1938, about one-half of one per cent were sterile, while surveys in the 1980s show an alarming increase to 10 per cent or more. Some of this could be related to high pollution levels, as could many bouts of disease that are written off as 'the flu'.

Here, as in many other cases of environmental problems, it is unlikely that science will be able to provide anything like definitive answers until it is too late. Society would be wise to make its decisions on first principles – subjecting people to contact with biologically active synthetic compounds should be minimised as far as possible. Cost-benefit analyses on the use of such synthetics should always be done with very conservative estimates of the possible health costs – that is, analyses should assume potential health costs to be in the high end of the range of uncertainty.

One of the most serious long-term problems of environmental contamination concerns a substance that is not at all toxic to people in low doses – carbon dioxide. The concentration in the atmosphere of this product of respiration and burning is steadily increasing, however, mainly because of two human activities: the burning of fossil fuels, and deforestation (and burning of the felled trees). It is estimated that the concentration of carbon dioxide in the atmosphere has risen from about 280 parts per million (ppm) in the early industrial age to about

350 ppm in the late 1980s – an increase of nearly 30 per cent. Worse, the concentration may double over pre-industrial levels by some time in the next century; how soon will depend on the rate at which fossil fuels are burned, and, perhaps, on how rapidly tropical forests are destroyed.

Unfortunately, carbon dioxide is not a simple emission problem, like other air pollutants. Carbon dioxide is an unavoidable product of combustion of carbon-containing materials such as fossil fuels and wood. The amount of carbon dioxide released is related to the amount of carbon in the original fuel; combustion of some high-quality coals (which are composed of more than 90 per cent carbon), for instance, may cause the release of three tons of carbon dioxide for every ton of coal burned. Proportionately smaller, but still substantial, amounts are released by the burning of petroleum, natural gas and wood.

Carbon dioxide, you may recall, is a greenhouse gas; it is transparent to the incoming short wavelengths of visible light in solar radiation, but absorbs and re-radiates the long wavelengths of infra-red radiation, which we sense as heat. As the concentration of carbon dioxide rises in the atmosphere, the average surface temperature of Earth will also increase. A doubling of the carbon dioxide concentration, it is estimated, would lead to a rise in the global average surface temperature of 3°C, plus or minus a degree or so. An average temperature increase of 1.5°C would make the planet's surface warmer than at any time in the last 100,000 years.

The inexorable increase in atmospheric carbon dioxide is not the only factor that is enhancing the greenhouse effect, however. Atmospheric concentrations of several other trace gases are also rising because of human activities. Some scientists think that their combined impacts may at least equal those of carbon dioxide, thus doubling the rate of atmospheric warming.

Methane is present at only about 1 per cent of the concentration of carbon dioxide, but it has an even stronger heat-trapping capacity. The atmospheric concentration of methane has been rising by about 2 per cent a year, released from rice paddies, flatulence of cattle and other ruminant animals, and from termites as they digest the felled trees in tropical forests. The processes are natural, but have been accelerated by expanded and intensified farming, herding and forest-clearing. Nitrous oxide, which is given off by the burning of nitrogen-rich fuels and by fertilisers and chlorofluorocarbons (CFCs) which are released from spray cans and refrigerants, are also significant contributors to the greenhouse effect, even though the quantities involved are by comparison minuscule. It has been estimated that the combined cumulative effects of all of these trace gases, including carbon dioxide,

127

could increase the global average temperature by the middle of the next century as much as 5°C.

A few degrees rise in temperature might seem like a small thing – perhaps an added discomfort for desert dwellers or a boon to air conditioner manufacturers. But the problem is not so simple. Increasing the average temperature of Earth could result in any of several changes in the circulation system, causing significant changes in climatic patterns, but a uniform temperature rise everywhere is an unlikely result. On average, surface temperature increases may be greatest in sub-polar regions and smallest in the tropics. But again, this may not be a consistent shift; weather patterns in any particular place can be expected to change in unpredictable ways. Some areas will get more rainfall and others less. Some places may be very much warmer, but others may cool as polar air moves more rapidly towards the equator. Perhaps more importantly, some climatologists predict a decline in soil moisture in the North Temperate zone.

Regardless of the direction of change, it will most likely be detrimental to agriculture in the short term; for, even where conditions 'improve', farmers will not be able to take rapid advantage of them. Newly moist semi-arid areas will not be quickly planted in crops, because it will be decades before the permanence of any changes will be clear. In the longer term, increased crop productivity due to the effect on photosynthesis of higher carbon dioxide levels might more than balance the loss of coastal farming areas that may be inundated by rises in sea level caused by the melting of polar ice caps. On the other hand, enhanced crop productivity is likely to be accompanied by increased populations of weeds and insect pests.

The disruption of agriculture and economic systems as climatic patterns change could potentially be an immense disaster for humankind. If the current build-up of trace gases continues, significantly altered climatic patterns can be expected to be clearly apparent by the early decades of the next century – a time when Earth will be even more overpopulated and the food situation even more marginal than now. As physicist John Holdren has written, unless some effort is made to change the trend, climate alterations induced by the buildup of atmospheric carbon dioxide and other trace gases could be large enough by 2020 or 2030 to generate a famine that might kill as many as a billion people.

Only one thing that human beings are now doing threatens a catastrophe more serious than the slowly developing threat from atmospheric carbon dioxide – or any other environmental threat. That, of course, is constructing ever more nuclear weapons. Throughout history, as advances in technological skills have given human beings ever greater power to obtain resources and turn them to supporting ever more people, parallel advances have also enormously increased the violence of warfare.

Wars have always caused environmental destruction, as well as killing people and shattering societies. Forests in ancient Greece were often burned in the course of battles, and the great trees of Britain were decimated by the Royal Navy's need for masts and spars during the era of wooden ships and iron men. In the landscape of places as different as the Argonne of France and New Guinea's Wewak, one can still detect the impact of this century's world wars. And extensive use of herbicides by US forces in Vietnam permanently altered the flora and fauna of that nation. Compared with such activities as growing crops and grazing cattle, though, the impact of the military on Earth's ecosystems so far has been minor and transient.

78. **Deformities from Agent Orange.** Herbicides used to defoliate forests in Vietnam so that enemy troops could be spotted from the air have caused lingering damage not only to the plant life of that country, but to human life as well. These children live in an area that was repeatedly sprayed with Agent Orange, which has been implicated as the cause of their birth defects.

But the explosion of the first atomic bomb at Alamagordo, New Mexico, in 1945 heralded the development of weapons that were potentially capable of destroying much of the planet. In the four decades since, the world's nuclear arsenal has increased from two weapons with a combined explosive power of roughly 35,000 tons of TNT (35 kilotons) to about 50,000 weapons with the combined explosive power of about 15,000,000,000 tons of TNT (15 million kilotons, or 15,000 megatons). To put this explosive power in perspective, if current nuclear arsenals existed in the form of Hiroshima-sized bombs, there would be enough to destroy a Hiroshima each second *for more than ten days*. That is why it is often said that the nuclear arsenals contain enough explosive power 'to make the rubble bounce.'

Military planners on both sides of the iron curtain are in agreement (privately if not officially) that any nuclear exchange would almost certainly escalate into a full-scale war. In such an event, the deaths of at least many hundreds of millions of people would be assured, along with a substantial portion of the cultural heritage of *Homo sapiens*.

Although many people seem to think that death in a nuclear conflict would come primarily in the form of vaporisation, the vast majority of the casualties would die in quite conventional ways, much like the victims of air raids in World War II. Men, women and children would be shredded by flying glass, fried by ignited fuels, crushed under smashed buildings, trapped in rubble, roasted to death by towering fires and so on.

On top of this would be added the horrors of radiation sickness — which can include intractable nausea and vomiting, diarrhoea, dehydration, haemorrhage, tremors, convulsions and (often even after an apparent remission of symptoms) death. And, of course, those 'lucky' enough to survive the radiation sickness would face a future of lowered resistance to infection and disease and an increased probability of cancer and having children with birth defects.

If the current arsenals were used in such a way as to cause the maximum civilian casualties worldwide (unlikely, but not impossible), some two billion people could succumb to blast, fire and the short-term effects of radiation. Should both the United States and the Soviet Union move, as is quite possible, into large-scale deployment of nuclear-tipped cruise missiles, the arsenals could rise toward 20,000–30,000 megatons, and the death of half or more of humanity from the direct effects would become much more likely.

Of course, the prompt deaths would just be the beginning of the casualty toll after a nuclear holocaust. Destruction of homes, hospitals, industrial plants, transport networks, power and water supply systems, sewage systems and so on would speed the dismemberment of society. Death rates from exposure, epidemics, starvation and banditry would undoubtedly skyrocket.

But not all the serious effects of an all-out nuclear war would be such direct ones. If large numbers of the ultimate pollutants, nuclear warheads, are loosed on the environment, they will do much more than just destroy people and their works. They, like many other forms of pollution, will also have a serious impact on Earth's life support systems. And it is those systems to which we now turn.

CHAPTER SIX

Disrupting the Biosphere

Over a few hundred thousand years, human beings have increased from a small, relatively restricted population in Africa to several billions spread throughout the land areas of Earth. While the proportion of land area directly dedicated to human habitation is small – perhaps 2 per cent at most – the impact of our presence is far more pervasive. Some 11 per cent of the world's land today is used to grow crops, and another 25 per cent is regularly grazed by domestic animals. Forests and woodlands cover about 30 per cent of the land area worldwide, and most of these are also exploited at one level or another by people. Thus human beings occupy or exploit in some degree over 60 per cent of the land surface of the planet. Most of the remaining area is under permanent ice or is desert, mountaintop, or too steep to be useful or accessible. And, although humanity's shadow darkens the land more deeply, the impact on the oceans is by no means negligible.

When this burgeoning population degrades the human environment, it also degrades the environment for most kinds of plants, animals and microbes. With the exceptions of domesticated creatures and opportunistic organisms that thrive in disturbed areas such as rats, cockroaches and crabgrass, virtually all of Earth's life-forms are suffering from the expansion of *Homo sapiens*. Almost all human activities, from farming, lumbering and mining to creating and releasing pollutants, make life difficult for the other organisms with which we share the planet. The biosphere – the part of Earth that supports life – is being altered in ways that threaten the extinction of a substantial proportion of biotic diversity.

What's wrong with that? After all, humanity has dominion over the entire globe; why shouldn't people turn its bounty entirely to serving human ends? One reason is that a fair number of people think it would be immoral to do so. If humanity co-opts 100 per cent of terrestrial productivity for itself and its domestic plants and animals, it will condemn to extermination millions of other species, from tigers and hawks to bees and wildflowers. Many people, including ourselves, are convinced that with dominion comes responsibility. Even if all those other species were perfectly 'useless' to society, we believe that, as Earth's stewards, people should preserve them. As far as anyone knows, they are humanity's only living companions in the universe.

Furthermore, we and many others also are convinced that other species are far from useless. To begin with, they benefit humanity because they are often beautiful and always interesting. What a dull world it would be without birds except for a few, such as house sparrows, starlings and Indian mynahs, that thrive in human-

79. Starlings. Starlings have been introduced in many parts of the world far from their origins. These opportunistic birds often displace native birds and can be pests. People have caused much environmental damage by introducing organisms into communities that have no evolutionary experience with them. The introduction of goats and mongooses on to islands, for instance, has many times caused ecological havoc.

80. Oryx. The Arabian oryx is a severely endangered species. Most of the few remaining live in nature reserves or in zoos.

dominated ecosystems! The billions of dollars spent annually by birdwatchers the world over in the pursuit of their hobby testifies to the aesthetic (and economic) value of birds. Would people enjoy a world without bison or butterflies, skunks or storks, oryx or orchids? The popularity of nature programmes on television and of nature books leads one to believe that a sizeable portion of the population would not.

81. **Whaling.** Many species of plants and animals have been endangered or pushed to extinction by over-exploitation. The whaling industry is still often guilty of overkill, and has brought nearly all the dozen or so species of great whales to the brink of extinction. This photograph shows pilot whales, which are not now endangered, being butchered in the Faroe Islands.

82. **Endangered butterfly.** This rare swallowtail butterfly was almost pushed to extinction in Britain, but is now hanging on in the Norfolk Broads, primarily in nature reserves. The disappearance of relatively inconspicuous organisms such as insects is seldom noted by humanity, but their loss is often a symptom of deterioration of the ecosystem in which they live.

On the other hand, a tour through central London, the United States Department of Agriculture, or the economics department of any major university could quickly convince a perceptive person that there are many people who couldn't care less if every organism not regularly served in fast-food restaurants disappeared immediately.

Appreciation of nature was once part of everyone's culture; a person understood the world around him or he did not survive. But today, apart from some Native Americans and a few remaining groups of hunter-gatherers in remote forests in the Amazon basin or the Philippines, nature appreciation is to a large degree an acquired taste. And many people, too poor or too culturally deprived, never acquire it. Even among citizens of the rich nations, an astonishing number of people who can differentiate a Porsche from a Rolls Royce without fail and know the names and faces of dozens of entertainers and sports stars have not the slightest clue about the difference between a sparrow and a flycatcher or why the leaves of plants are green.

83. **Black rhinoceros.** These three black rhinos were photographed in Ngorongoro Crater, Tanzania, a national park where a few protected individuals still survive. African rhino populations have been reduced by 90 per cent or more in the last twenty years by hunters and poachers. Rhinoceros horn is prized by some Asian cultures for its supposed aphrodisiac qualities and by some Middle Eastern societies as the raw material for daggers used in puberty rites.

Interest in nature among some such people picks up when they are told that humanity gains vast direct economic benefits from other organisms – that, for example, active ingredients in a third of all prescribed medicines are chemical compounds found in wild plants. Moreover, many other medicinal compounds have been designed by chemists to mimic natural plant compounds. Remember, plants long ago developed the ability to poison their enemies or otherwise interfere chemically with their feeding. The chemical armamentarium that plants have evolved includes substances as diverse as quinine, digitalis, morphine, caffeine, spices, rubber and many components of petroleum.

Few people, indeed, realise that humanity has drawn the very basis of its civilisation from the natural world; that, in fact, without the descendants of three unspectacular species of wild grasses, now known as wheat, maize and rice, there would be no United Kingdom, United States or Union of Soviet Socialist Republics.

Fewer still know that humanity has barely scratched the surface of the riches nature has to offer; that the relative handful of plants and animals that have benefited *Homo sapiens* directly could be only the beginning. Every time a square mile of Amazonian rainforest is cleared, a possible cancer cure or basis for gasoline farming may be lost forever as an obscure (to us) plant population or species goes extinct. Many presently unexploited plants have obvious potential for development into useful crops. Only some 150 species of higher plants (out of perhaps a quarter of a million species) have been grown commercially, fewer than twenty species are important in the human diet, and those three grasses just mentioned really make up the feeding base of our species.

A United States National Academy of Sciences report put it very succinctly: 'These plants are the main bulwark between mankind and starvation. It is a very small bastion.' At the moment, the possibility still exists for greatly expanding that bastion, but the very increase in human numbers is destroying that potential. On the one hand, the desperate need to increase food production every year to support ever-growing populations has resulted in the wide dissemination of a relatively few high-yielding strains of the major crops. These new varieties, although very productive when properly cultivated, are displacing a multitude of traditional strains of the same crops. As a result, the genetic base of the crops is being narrowed, and that in turn reduces the ability of plant geneticists to select new strains to cope with new strains of pests or changing climatic conditions.

At the same time, the expansion of the human population is also leading to the extermination of wild plant species and populations around the world. Some of the plant populations being lost are wild relatives of the domestic crops, which, like the traditional strains of the crops, are potential sources of valuable genetic information that could be incorporated into the crops by breeding programmes. Furthermore, an unknown number of the disappearing plant species might, if preserved, provide the basis for developing entirely new and extremely valuable crops.

For example, a group of tropical plants known as grain amaranths contain high-quality protein in their grain, and even their leaves are protein-rich (and are widely eaten as a kind of tropical spinach). One or more of the amaranths could well be developed into staple crops, substituting for wheat (which does not grow well in the tropics) and maize (which is nutritionally inferior).

The potential benefits to be gained from growing new protein-rich crops in the hungry tropics are incalculable. But to turn this potential into reality would require a determined programme both to develop strains suitable for widespread commercial and subsistence cultivation

135

and to overcome cultural prejudices that favour traditional crops. And, of course, it is essential that the diversity of strains and species of amaranths – cultivated or wild – now growing be preserved, for they are the raw materials from which geneticists could produce new crops. Once the genetic diversity of amaranths is reduced too far, the opportunity will be lost forever.

84. **An open market in Burkina Faso.** Not only are wild plants and animals an invaluable source of potential new foods, but so are the plants cultivated in traditional societies, which until recently have been neglected by plant breeders and agronomists. Many of the vegetables and fruits sold at this country market in West Africa would be unfamiliar to most Westerners. Unfortunately, many traditional crops, which might help to improve diets of poor people in many developing nations, are being lost or genetically impoverished as farmers switch to growing only the improved high-yield strains of the major grain crops. The neglected traditional strains are endangered species, too, and of immeasurable importance to humanity.

There are many other examples of known potential among wild plants, animals and micro-organisms for producing new and extremely valuable crops, domestic animals, medicines, industrial products and fuels. But these, of course, are just potentials. And all of us are now living in the high discount world of contemporary economists, a world dominated by a short-sighted attitude of 'use it up today and enjoy the profits, for tomorrow it will be less valuable'.

Thus the potential values to be found in the world of other organisms are essentially non-existent to conventional economists – and to businesspeople, politicians and others who share their myopic view of reality. They would recommend turning the Amazon basin into a parking lot if the present value of the parking fees were greater than that computed for the possible cancer cures or life-saving new crops that would be lost forever by clearing the species-rich forests. Whoever said that economists know the cost of everything and the value of nothing was close to the mark. But it turns out that they often have a view of costs that is as naïve as their views of value are misguided.

Nowhere is this clearer than in the attitudes of conventional economists and their followers towards the mounting extinction epidemic. They have decided that the loss of diversity has been exaggerated (which is certainly wrong) and that what has been lost so far is not as valuable as the economic benefits gained in the process of destruction (quite likely wrong). Confident of these truths, they assume that continued economic growth will be beneficial in perpetuity (assuming the preposterous is a cornerstone of economics). But the overwhelming cost that civilisation will pay for the extinction epidemic has not even entered into the thinking – or the calculus – of mainline economics.

85. A burned tropical forest. Forests are disappearing rapidly throughout the tropic zones. By one estimate the world's tropical moist forests – the type of ecosystem that harbours the greatest diversity of life – are being destroyed or degraded at a rate of roughly 200,000 square kilometres per year. Thus an area of tropical moist forest equivalent to the land area of the United Kingdom is vanishing every fourteen months. The patch of Malaysian forest in this photograph has been cut and burned for conversion to agriculture, the prime cause of disappearing tropical forests. The forests of the Malay peninsula may well be entirely gone by the 1990s.

That cost is the loss of the irreplaceable life-support services that natural ecosystems supply gratis to civilisation. These services include regulation of the quality of the atmosphere, amelioration of the climate, provision of fresh water, disposal of wastes, recycling of nutrients, generation and maintenance of soils, control of the vast majority of potential pests of crops and carriers of disease, provision of food from the sea, and the maintenance of a vast 'genetic library' of wild populations and species from which humanity has already drawn – and can continue to draw – enormous benefits.

137

A full complement of the other living organisms of Earth is essential, along with Earth's physical environment, for the provision of these essential services. In most cases, humanity has no idea how to provide the services should ecosystems cease rendering them. And where humanity does have an idea, it is clearly beyond its capability to provide them on the scale required. In short, by altering the biosphere in ways that could exterminate most of its fellow passengers on Spaceship Earth, *Homo sapiens* is quite rapidly sawing off the limb on which it is perched. The ultimate cost of the extinction epidemic, the one thus far overlooked by the economists, may be the collapse of civilisation.

Let us review these vital services and consider how they are threatened by civilisation. First, recall that the composition of the atmosphere, which is largely a result of aeons of coevolution with life and still partly controlled by the activities of organisms, can be changed by human actions. Fortunately, the atmospheric pools of the two principal components, nitrogen and oxygen, are so vast that rapid changes in them are highly unlikely. While the proportion of oxygen is critically important to life (too little would lead to oxygen starvation of all respiring organisms; too much might ignite even the dampest forests), thousands of years would probably be required for substantial depletion of the oxygen pool, even if today's ecosystems were virtually destroyed. This should provide little solace, though, for if those systems are destroyed, *Homo sapiens* will go with them, and we won't have to worry about suffocating or burning up.

The concentrations of other atmospheric gases are not so well buffered against change. The thickness of the ozone shield that permitted life to leave the seas and invade the land is in part controlled by biological processes. The same can be said of the tiny proportion of carbon dioxide in the atmosphere, which, as we have seen, plays an important role in regulating climate (which, in turn, is crucial to maintaining agricultural productivity). Dust and other pollutants are also filtered out of the atmosphere by vegetation, especially forests.

Besides their influence on the atmosphere's composition, natural ecosystems help to ameliorate climate and weather, on both global and local levels, by influencing patterns of atmospheric heating and air flow. Ecosystems often modify the albedo – the proportion of sunlight reaching the surface that is reflected – and thus how much the surface is heated by the sun and how much cloud cover is formed. Remember, the characteristic cloud cover of the Amazon basin in Brazil is generated in large part by the tropical forest recycling rainwater numerous times as the moisture-laden air passes over it. If the forest were completely removed, many of the big rain clouds would vanish with it, reducing the albedo of the entire area, and it would desertify (as has north-eastern Brazil, which was deforested in the last century).

No one knows just what effect deforestation of the Amazon basin would have on climate elsewhere, because the details of the climatic system are not well enough understood. But it could be significant, especially since it *is* known that relatively small perturbations (such as a small increase in the tiny proportion of carbon dioxide in the atmosphere) can cause far-reaching changes. As atmospheric physicists sometimes say, the global climatic system is governed by small differences between large numbers.

Regardless of the possible distant effects, deforestation of the basin would not only wipe out one of Earth's greatest reservoirs of organic diversity, it would also completely thwart Brazil's ambitious development plans by destroying the potential economic value of that prime resource and much else besides. The native Amazonian ecosystem's amelioration of the climate of the area must be maintained if the region is to be productive for human beings.

In a very different region — the Sahel — it has been proposed that overgrazing has contributed to desertification by influencing the albedo. The theory goes like this: denuded soils have a higher albedo than the original grass-covered surface; removal of the grass led to less heating of air near the surface (since less solar energy was absorbed by the surface), fewer updrafts and thus less cloud formation and rainfall. On a local level, of course, the impact of ecosystems on climate is familiar to any farmer whose soil has benefited from the windbreak effect of a nearby forest or any hiker who has sheltered in that cool forest on a hot day.

The role of forest and grassland ecosystems in maintaining the cycle and flows of Earth's fresh water benefits humanity in a number of ways, apart from its profound influence on climate. The benefits, unfortunately, more often come to be appreciated when they have been lost.

If the plants of a forest or grassland are removed, for instance, nothing stops the rainwater from running swiftly downhill over the land's surface. With no plant cover, soil is no longer held in place, and it is washed away with the water to collect in streams, lakes and bays. Not infrequently, this accumulation of silt clogs irrigation ditches and shipping channels and fills up reservoirs behind dams, soon rendering them useless. The process of aquifer recharge may also be disrupted by water running off the surface unhindered rather than sinking into the ground. Meanwhile, the continual loss of topsoil to erosion gradually diminishes the land's productivity.

The climatic result of the rapid run-off of rainfall is an accentuated alternation of local floods and droughts, which can also reduce productivity. Rather than the water being metered out in a steady flow

by a well-vegetated watershed, it runs swiftly off the land in swollen, silt-laden streams after each rainfall, and the land quickly dries out. The greater the area bereft of vegetation, generally, the more widespread and severe may be the flood/drought cycles.

In the poverty-stricken East African nation of Rwanda, the Parc National des Volcans containing the Virunga volcanoes encompasses less than 0.5 per cent of the country's land surface. The unbroken forests of the park act as a gigantic sponge, soaking up the rains that fall on the mountains and slowly releasing the water to the lowlands in steadily flowing streams. The Virungas supply about 10 per cent of Rwanda's agricultural water — and they also provide sanctuary for about 240 mountain gorillas, among the very last of their species.

Pressures on the park land are enormous, as Rwanda's human population is growing fast enough to double in just 19 years. Less than a decade ago, the park was nearly twice as large, but 40 per cent of it was cleared for an ill-fated scheme for growing pyrethrum (a natural insecticide) as a cash crop. One result was the disappearance of several local streams.

Rwandans are desperate for land; a man has no status without it. But outside of the Virunga range and another small national park, the countryside is already cultivated right to the mountaintops. Small wonder that farmers are creeping into the park, which has neither fences nor permanent boundary markers, a few feet at a time. If the entire park were cleared for agriculture, it would buy Rwanda about six months' breathing space at current population growth rates — at the cost of an agricultural disaster caused by disruption of the water-management services of the Virunga ecosystem.

Not long ago, it was discovered that the annual flood of the Amazon River at Iquitos, Peru, had gradually been increasing in volume since about 1970 as a result of deforestation around the river's headwaters in Ecuador and Peru. Similarly, deforestation in the Himalayas of northern India and Nepal has produced catastrophic flooding along the Ganges. Indeed, almost anywhere one travels in the poorer parts of Earth, the rivers are running brown with silt, which is indicative of massive deforestation and an absence of soil conservation practices.

Not that things are so much better in rich nations. The water and soil management functions of natural ecosystems have been badly damaged in places as diverse as the Mediterranean basin and the American Midwest. Much of the damage in the Mediterranean was done long ago, and that once heavily forested and well-watered land is now an ecological disaster area. A trip through southern Italy or the Peloponnesus reveals to those who can read the message how ecological mismanagement can bring down prosperous world powers.

Similarly, much of the south-western United States was once not

86. Slash and burn agriculture. In the past, slash and burn agriculture, in which a farmer cleared a small patch of forest, cultivated it for a few years, then moved on to clear another patch, allowing the forest to regenerate, was a sustainable way of life and inflicted no permanent damage on the forest. Today, rapidly expanding populations in the tropics cause farmers to return to old patches before recovery has been completed and to make bigger clearings (to compensate for lower soil fertility). The result is progressive degradation of the forest and it is a principal means by which tropical forests in many regions are being destroyed. This clearing is in Papua New Guinea, which has not yet been heavily damaged in this way.

87. Logging in a tropical forest. Much loss of tropical forest is caused by logging, whether clear-cutting or selective logging. Even when selected trees are felled in a tropical forest, many others may be injured or killed because of the tangle of vines connecting them, by being knocked over by a felled neighbour, or by the building of logging roads. And the loggers are usually soon followed by farmers and fuelwood gatherers. Commercial logging affects some 45,000 square kilometres of tropical forest each year. This large-scale operation extracting timber for export is in Kalimantan, Borneo. Much of the pressure to log tropical forests is generated by demand for lumber and pulpwood (for paper and cardboard manufacturing) in rich countries.

88. Fuelwood gathering. A third major cause of destruction of tropical forests, and even more of drier scrub forests and savanna trees, is the cutting and gathering of fuelwood by people in increasingly overpopulated developing nations. Wood is the principal source of energy for most people in poor countries, where women may spend large amounts of their time and energy finding and harvesting it. In many regions, the wood harvesting exceeds rates of regrowth, sometimes by many times, and the scarcity of wood for such essential uses as cooking is acute. Fuelwood harvesting is also an important contributor to desertification. These women are bringing home logs from the Kakamega forest in Kenya.

desert but grassland. The arid scrub communities that occupy the area today are in large part a product of gross misuse by farmers and grazers – misuse that continues today under the auspices of the Bureau of Land Management and the US Forest Service's famous 'land of many abuses' policies.

In the rich central prairies of North America, things are moving in the same general direction. The process of destruction is slower, however, because the soils inherited by European settlers there were among the deepest and richest in the world. Nevertheless, it is estimated that the average depth of the soil has been reduced by half or more since the first ploughs penetrated the surface less than 150 years ago.

Soil, like fresh water, *should* be a renewable resource, since it is continuously produced and enriched by natural ecosystems. If soils were only pulverised rock, then soil ecosystems and their services would be relatively easy to maintain or replace. It is the crucial and fragile organic components – the myriads of tiny soil organisms and the nutrients they continuously recycle – that are so readily lost or destroyed. Left to its own devices or properly husbanded, the soil ecosystem maintains itself.

Unfortunately, in far too many parts of the world, soil ecosystems are neither being left to their own devices nor husbanded. Instead, they are being treated as a non-renewable resource, managed so that in many parts of the world, erosion is carrying them away at rates of inches per decade. Agricultural economist Lester Brown, president of Worldwatch Institute, has explicitly warned that while civilisation might well survive the depletion of fossil fuels, it cannot survive the destruction of Earth's topsoil. He and his colleagues have roughly estimated that, worldwide, perhaps 23 billion tons of soil, in excess of that replenished by natural processes, are washed off the land every year. One to two billion tons of soil are eroded annually from farmland in the United States above 'tolerance' levels – that is, above levels of loss that supposedly will not reduce productivity in the long term. Such excess losses are occurring on more than a third of the nation's cropland. The situation is, if anything, worse in many other major agricultural regions.

Changes in farming methods, especially to continuous cropping rather than crop rotation, account for much of the acceleration in erosion rates in the past generation, although consequent declines in productivity may be masked for a time by increased applications of the fertilisers that allow continuous cropping. But fertiliser applications cannot replace soil; sooner or later, continued soil loss results in a serious loss of productivity – a drop of approximately 6 per cent for each inch lost from moderately deep soils. Once erosion has proceeded

143

far enough, the damage is for practical purposes irreversible. And indeed, ominous declines in productivity have been seen in many areas of the world.

89. Forest mismanagement, Western Australia. Much of the remaining natural forest in the world is being lost through over-exploitation and mismanagement. This patch of jarrah forest, unique to Western Australia and now extremely rare, has been severely degraded through poor management practices. In this case, the forest has been subjected to 'controlled burning' every five years to reduce undergrowth and prevent wildfires. Much less frequent burning would probably serve the purpose without unduly damaging the forest, especially since the townships once threatened by wildfires no longer exist.

Another important service supplied by ecosystems is waste disposal. It is, of course, convenient to have garbage and dead bodies disposed of. It is nice that the corpses of cats squashed on highways disappear as magpies, maggots and bacteria do their jobs. The organisms that carry out decomposition do their jobs so well that people have recruited some of them – bacteria – to perform the same functions in sophisticated human-designed sewage treatment systems. But the main importance of decomposers is not aesthetic. Were it not for them, Earth's great nutrient cycles would grind to a halt, and we would all die.

Human dependence on plants to eat, directly or indirectly, is shared by myriads of organisms, including all animals. It makes life difficult for the plants, and they have responded (as noted earlier) by becoming poisonous, difficult to digest, or otherwise unpleasant to eat. But, quite naturally, when human beings develop crops by selective breeding, they tend to choose the least poisonous, most digestible, most readily eaten individual plants to be the parents of the next generation. The process of domestication and crop improvement is generally one of producing plants genetically designed to be both relatively nutritious and relatively defenceless. In the process, of course, humanity is not just providing itself with nourishment, it is spreading the table for a vast diversity of competitors.

Ever since human beings began to grow and store crops, they have had to share them with other organisms, especially insects. But, considering the richness of the resource and the plethora of other plant-eaters, it is really quite amazing that so few other creatures have joined the corn earworms and flour beetles to become serious competitors for the bounteous human food supply. For this we can thank one more free service provided by ecosystems: pest control.

The overwhelming majority (over 97 per cent) of herbivorous insects and other creatures that could potentially become serious pests of crops never do so, because their populations are controlled by natural predators. This is demonstrated every time humanity intervenes in ecosystems in a way that diminishes or destroys populations of those predators. Sometimes the human error is obvious. Instead of using time-honoured herding procedures to protect their flocks from the depredations of coyotes, many stockmen insist that the government not only lease them public land for grazing at prices that amount to a gigantic subsidy, but that it destroy the coyotes to boot.

Stockmen as a group are, in this respect, advocates of capitalism for everyone else and socialism for themselves. When the government has some rare success in reducing the coyote population, it often has to take on rodent control as well, because coyotes eat many more rodents than lambs. Killing predators frequently leads to population explosions in their prey.

More often, the human error is a little more subtle. Pesticide companies love to recommend repeated and widespread application of their products. This approach virtually guarantees that the pests will evolve resistance to the poison, just as the peppered moths evolved melanism in response to the change in *their* environment. The insecticide company is delighted to have gardeners spray their cabbage patches several times a year with their latest bug killer, ZAP. Ten or twenty generations of surviving and breeding by individuals that happen to have more resistance to ZAP than their cohorts produces a population of insects that enjoys ZAP as an aperitif before devouring the cabbage.

When a US-based chemical company was informed by scientists a few decades ago that one of their pesticides was making pest problems worse for cotton growers, they advertised: 'Even if an overpowering migration [*sic*] develops, the flexibility of [the pesticide] lets you regain control fast. Just increase the dosage according to label recommendations.' What could be better than a product that requires more and more to be used if it is to keep on working? Many farmers have, in fact, been hooked by the pesticide manufacturers, abetted by state and federal agencies, on incompetent pest control regimes in a manner analogous to the way drug dealers hook schoolchildren.

But the development of resistance is only one consequence of the misuse of pesticides. Another is the 'promotion' of previously harmless organisms to pest status because the pesticides interfere with the pest control service of the ecosystem. A classic case occurred in the Cañete valley of Peru, an important cotton-growing area. Shortly after World War II, growers there were pleased to discover that treatment with new synthetic pesticides greatly increased their harvests. They decided that, if a little DDT was a good thing, a lot would be even better. Spraying was stepped up, and traditional pest control measures neglected. The result was, first, an increase in cotton yields, followed quickly by a disastrous plunge to below pre-DDT levels. Not only did the original pest populations become resistant to DDT and explode to new levels, but several *brand new pests* put in an appearance.

The old pests resurged and new ones turned up chiefly because populations of predators are almost always more susceptible to pesticides than are the herbivorous pests that they eat. One reason is that pest populations are normally larger than those of their predators (they feed lower on food chains and therefore more energy is available to them). Purely as a matter of chance, it is less likely that all the individuals in a large population will be exterminated by a broadcast poison than that all those in a small one will. Some, for instance, may just happen to be sheltered from the spray or outside the area being treated. A larger population is also more likely to include genetic variants (again by chance) with some resistance to the pesticide. These surviving resistant individuals can begin the process of evolution of resistance in the entire pest population.

Moreover, pest organisms are more likely than predators to have some resistance to poisoning in general even before they are exposed to the pesticide. For millions of years, herbivores have been coevolving with plants that have been trying to poison them in various ways. Predators, on the other hand, have relatively little evolutionary experience with poisonous chemicals. Thus an attack with a pesticide is not as novel an experience for a herbivore as it is for a predator.

When the natural pest control services of ecosystems are disrupted, the usual result is a loss for humanity, except for the pesticide pushers. When DDT killed off the insect predators of spider mites, those small relatives of ticks (which previously virtually never did serious damage to crops) became a major category of pest. The pesticide industry was unfazed; it just started manufacturing extremely toxic and often carcinogenic 'miticides' to wipe out their own creations!

Thus a basic problem with pesticides is that they tend to be more effective against friends than foes, more readily killing predacious insects, birds and other enemies of pests than the pests themselves. In addition, one group of pesticides, insecticides, often wreaks havoc with

90. **The Dust Bowl.** In the 1930s, over-cultivation and over-grazing in the central and western United States combined with several years of drought to create the famous 'Dust Bowl'. A dust storm in Idaho buried this car in March 1937. Following those disastrous years, the United States Agricultural Department instituted effective measures to control wind-caused soil erosion and prevent a repetition of the Dust Bowl. Unfortunately, in recent years, pressure to produce ever more grain for the competitive world market has led to widespread abandonment of erosion control practices.

91. **Ethiopian dust storm.** Dust storms are not a thing of the past, as this photograph taken in Ethiopia in the early 1980s attests. The process of desertification, especially removal of the vegetation, has led to this extreme situation. The vegetation has been removed mainly by over-grazing and the cutting of wood for fuel.

another free ecosystem service – the pollination of crops – by killing bees and other pollinators. In the United States alone, insect pollination is essential for some 90 crops and benefits several others.

Of course, natural ecosystems do not just help provide humanity with food by maintaining nutrient cycles and controlling floods, droughts and pests in support of the agricultural systems that are embedded in them. They also provide nearly all the food from the sea (mariculture makes a minor contribution) consumed by human beings as well as a significant amount of food from the land, such as game, wild berries and nuts. Those few societies that still hunt and gather are of course completely dependent on natural systems for their food and other needs. And natural ecosystems maintain the genetic library that in the past made civilisation possible, and which holds so much promise for aiding humanity in the future.

In summary, when *Homo sapiens* assaults other species of organisms, it is assaulting parts of its own life-support system. Dealing with the assaults is difficult because they are so varied and so closely tied into many, if not most, human activities. Extinctions occur when areas are paved over for urban development or ploughed under for conversion to farmland. Extinctions occur when forests are soused with acid precipitation, or when toxic chemicals are dumped into lakes, rivers and streams. Extinctions are caused by over-grazing, over-exploitation of commercially valuable organisms, disposal of tailings from mining, and climatic change. They also result from the transport of animals from one place to another: goats have wiped out many plant populations on islands, and mongooses introduced for rat control have done the same for many bird populations. And, of course, if there should be a large-scale nuclear war followed by a nuclear winter, there could be an extinction event of the same magnitude as the one 65 million years ago that ended the hegemony of the dinosaurs.

92. **A stream protected by vegetation.** This well-vegetated mountain stream in Catoctin Mountain Park, Maryland, remains clean and clear and does not dry up between rains.

93. **Eroded slopes in Haiti.** Widespread deforestation in mountainous Haiti has led to massive losses of topsoil from the mountain slopes and contributes to Haiti's poverty as farms quickly lose productivity.

94. **Haiti's silt washing into the sea.** Topsoil washed from the denuded mountain slopes by heavy tropical rains clogs rivers and runs into the ocean, where it can damage offshore ecosystems, including fisheries that could help to nourish Haiti's hungry population.

To some degree, such assaults seem inevitable; acid precipitation might be abated, but some climatic change from the build-up of carbon dioxide and trace gases in the atmosphere is essentially unavoidable. Pollution can be contained, but humanity still needs the minerals extracted from mines. People might greatly restrict the inadvertent transport of organisms, but the demands of a burgeoning human population for protein will make it extremely difficult to stop the over-exploitation of fish stocks. And the day-to-day acts involved in what is thought of as normal living add to the problem.

Every time a new motorway is built through the English countryside, every time a new home site is carved out of California chaparral, a new bit of semi-arid desert in Kenya is brought under the plough, or a new patch of Amazonian forest is cleared for an oil drilling pad, populations (or even species) of other organisms are forced into extinction. But can ecologists prove that such 'minor' acts seriously endanger the human future? What harm can be done by one more motel, dam, farm, motorway, plantation or what have you? Can ecologists give any advice on what limits should be placed on acts that reduce organic diversity?

Sadly, in the vast majority of cases, ecologists cannot accurately predict the sequence of events that would follow any given extinction, any more than political scientists can predict the consequences of the death of a single politician or an economist can trace out the consequences of the failure of a single business because of foreign competition. Such single events as the loss of one population or species in an ecosystem, or of one politician or business in a society, may have the potential to cause serious dislocations, but in most instances they don't.

The inability to foretell the exact consequences of a single event nevertheless does not prevent easy prediction of the results of a large number of similar events. Ecologists can predict pretty well what might happen if a great many populations or species were exterminated, just as a political scientist could probably predict the consequences of the assassination of an entire government or an economist the failure of a large number of businesses.

Ecosystems are always undergoing changes, and populations and species not uncommonly go extinct. But ecosystems are normally under 'progressive maintenance', as are the aircraft of responsible airlines. Over the reaches of geological time, extinctions have been more than balanced by the formation of new populations and species, just as lost or damaged parts are always being replaced and malfunctions corrected in aircraft. And, just as only someone who was insane would knowingly patronise an airline with no maintenance programme, so only a madman would want to ride on Spaceship Earth

95. **Parc des Volcans, Rwanda.** One of Africa's most densely populated nations, Rwanda sacrificed some 40 per cent of its national park in the Virunga volcanoes during the 1970s in order to develop pyrethrum plantations. The daisy-like pyrethrum flowers can be seen in the foreground here. The forested slopes of the park supply some 10 per cent of the nation's agricultural water; but, after the forest was removed, several streams disappeared and the area's rainfall became less dependable. The pyrethrum scheme, moreover, has been an economic failure.

96. **Silverback gorilla.** This magnificent male mountain gorilla is among the last of his gentle kind, one of only about 240 individuals surviving in the forests of the Virunga volcanoes that straddle the borders of Rwanda, Uganda and Zaire in central Africa. Lowland gorillas, a closely related subspecies, are more abundant, but their future nonetheless seems bleak.

151

if the components of its ecosystems were being dismantled so fast that maintenance could not begin to keep up repairs. Yet here we are, with no other spaceline offering transport.

But, even beyond their crucial importance in preserving the planet's habitability, there is still another reason to preserve and protect the integrity of the biosphere. Human beings, like virtually all organisms, are completely dependent on the energy produced in photosynthesis for food. People also use a good many other products derived from living systems. But the amount of photosynthesis that occurs on Earth, though vast, is distinctly limited. The ultimate constraint on human population expansion, then, is the amount of organic material that can be generated annually by Earth's producers on land and in the oceans.

Before the most recent surge in human population growth following World War II, photosynthesis on land produced perhaps 150 billion tons of dry weight of organic matter each year. Now, thanks to the activities of our species, the annual production of organic material in terrestrial ecosystems (both natural and human-controlled) has fallen to only about 130 billion tons. Humanity is thus responsible for a net decrease in global photosynthesis – a diminution in what may be thought of as the basic food production for all of Earth's organisms.

Some of the reasons for the decline in productivity are fairly simple and obvious: photosynthesis cannot occur on or under buildings, parking lots, airports, streets or highways. But more subtle land-use changes in many regions have also contributed to reductions in photosynthesis. For example, huge areas of the American Southwest, which have been desertified by overgrazing, were once productive, though arid, grasslands. The same is true for other vast regions on virtually every continent of semi-arid and arid land that have been desertified under human impact. The world's forested areas, the most productive per hectare of all ecosystems, have been substantially reduced in extent in this century, mostly for agriculture. In most cases, significantly less photosynthesis is carried out in agricultural ecosystems than took place in the natural ecosystems they replace.

In addition, though, burgeoning numbers of people have been co-opting an ever-growing share of that total production. About 3 per cent of global net primary production (4 per cent on land, 2 per cent of oceanic production) is consumed directly by people and their domestic animals: eaten, burned as fuel, used for construction (wood) or clothing (cotton) etc. This seems like a relatively small amount – until you consider that *Homo sapiens* is but one of somewhere between five and thirty million species of consumer organisms that must share Earth's bounty. In that light, even a half of one per cent might be excessive.

But direct consumption is only the beginning. As noted earlier, vast areas of the planet have been converted to simplified agricultural systems, from which most kinds of plants and animals found in local natural ecosystems are excluded. All the photosynthetic production in those systems should be chalked up to humanity's account. So should the production in forest plantations and the products of photosynthesis destroyed but not used when forests are logged, as well as those consumed in human-set forest and grass fires. Such calculations reveal that people are co-opting about 30 per cent of terrestrial production.

Even this figure is an underestimate, however. It does not include the potential production lost because of paving over habitat, desertification, deforestation and conversion to less productive systems mentioned above, which in a scant four decades has amounted to about 13 per cent of the estimated earlier production on land. When those losses are factored in, it turns out that our one species has co-opted or destroyed some *40 per cent* of potential terrestrial productivity. Since the fraction co-opted in the oceans – only about 2 per cent – is much smaller than that on land, about a quarter of all of Earth's photosynthetic production, land and sea, is appropriated by people.

But we should not take heart from the 'limited' human exploitation of the abundance of the oceans. All the evidence indicates that it would be very difficult to increase very much the harvest of food from

97. **Gulley erosion.** Over-grazing and deforestation have combined to produce this spectacular erosion in the Kiritiri area of Kenya.

153

the sea. Humanity feeds high on oceanic food chains, where the laws of biology and physics dictate that the available energy is far, far less than was originally captured by photosynthesis.

There are not all that many fish in the sea, after all. Fisheries production has grown much more slowly than the population since about 1970, and what growth there has been has come primarily from exploiting less desirable stocks as the more desirable ones have declined from overfishing, pollution and disruption of habitat. It would be exceedingly difficult to increase much more the harvest of fish from the sea; doubling it may prove impossible.

Population and photosynthetic production numbers can tell us a great deal about the human future. On land, one species is now using or reallocating a very substantial portion of the planet's 'income', and in all likelihood it is demographically committed to doubling its population. Most of the production to support such an expansion will be sought on land, where nearly half the photosynthetic income has already been taken or foreclosed. Since production per person of oceanic fisheries has been dropping, people will have to increase their use of terrestrial productivity simply to compensate for the loss of food from the sea.

On land, however, as we have seen, human activities have been steadily eroding productivity. In most regions, the conversion to agriculture of significantly more land seems unlikely in the near future, and much of the land now in use is losing productivity through soil erosion and nutrient exhaustion. Manufactured fertilisers cannot fully compensate this loss; and once the organically rich upper layer of soil is destroyed, no amount of the standard fertilisers can produce crops.

In addition to increased production of food and fibre (including timber) that will be needed by the expanding population, moreover, there has been much talk of using energy from biomass (either firewood or fuels derived from plants grown for the purpose) to substitute for some of the energy from declining stocks of fossil fuels. That is, the plan is not only to continue increasing the usual demands on humanity's income, but to use some of the income as a substitute for using our dwindling capital. While returning to dependence on income is generally the wisest course for the long term, this move would inevitably further intensify the mounting pressures on the planet's limited sun-derived income and could accelerate the exhaustion of soil and groundwater capital.

Thus it would appear that *Homo sapiens* is attempting to take over *all* Earth's terrestrial production to support its population – even without 'planning' for any increase in the average person's standard of living. We will leave to your imagination, for the moment, the likelihood of success, given the importance of all the other organisms in operating the biosphere and maintaining the habitability of Planet Earth.

PART THREE

The Human Response

CHAPTER SEVEN

The Coming of the Green

That humanity can have a heavy and deleterious impact on Earth is, in itself, nothing new. But it was not until this century that the collective impacts of human societies reached a global scale, threatening a nearly simultaneous collapse in all important regions of the planet. Fortunately, however, the last part of this century has also produced an unprecedented recognition of the causes and consequences of those impacts and an increasing urge to seek remedies. This greater awareness has none the less been achieved through long and sometimes painful experience.

The first known case of ecological collapse of a civilisation occurred several thousand years ago in the valley of the Tigris and Euphrates rivers in what is now Iraq. The Mesopotamian civilisation which had arisen there was 'hydraulic' – utterly dependent on irrigation from those two great rivers. With an assured water supply and the invention of the plough, farmers could grow much more food than they needed for their own families, and the availability of surplus grain opened the door to the development of cities.

But irrigation is commonly a temporary game: dams and canals silt up, and land may become infertile because of waterlogging or salt accumulation. Even societies with the most modern technology are hard pressed to prevent such deterioration, as farmers in California's Imperial Valley know only too well. The Mesopotamians never had a chance. With inadequate technology and frequent harassment by invaders, they suffered the first great ecocatastrophe, and their civilisation gradually collapsed.

The Greeks and Romans were not much more successful. When the civilisation at the eastern end of the fertile crescent had faded, the Mediterranean Basin was still a relatively well-watered land, mostly covered with thick forests. For instance, Corsica had tall trees so crowding its shores that the masts of the ships of the first settlers were smashed by giant branches extending out over the sea. And those Mediterranean forests covered rich soils that would one day support the granary of the vast Roman Empire.

The first people to make changes were the ancient Greeks. Their land was carpeted with rich stands of pines, oaks and other trees with thick, drought-resistant leaves – what plant ecologists call 'sclerophyllous forest'. But expansion of the Greek population led to the progressive destruction of the forests to satisfy the need for lumber, firewood and charcoal (used in firing pottery and in other industrial processes). And, in a process still going forward today in places as disparate as Amazonia and Australia, forests were destroyed simply to create more pasture.

Regeneration of the Greek forests was prevented by a combination of severe soil erosion and goats. Those 'horned locusts' are engines of ecological destruction wherever people have taken them, wiping out all but the toughest and least accessible vegetation as well as other, less hardy animals dependent on that disappearing plant life. Goats have left their ruinous mark over much of Earth, but the 'goatscape' they helped create in the Mediterranean basin is their best-known monument.

98. **Goats as agents of desertification.** Uncontrolled grazing and browsing by goats has contributed to desertification in wide areas of the world, including the Mediterranean basin and much of Africa. These goats are attacking a struggling thorn tree in a desertified area of Niger. The tree will nevertheless survive unless the herdsman cuts it down to give the goats access to all of its leaves.

It is a monument that could have been appreciated by many Greek intellectuals, who understood early on what was happening to their land under the impact of human activities. Plato, writing four centuries before Christ, said of the region around Athens: 'What now remains compared with what then existed is like the skeleton of a sick man, all the fat and soft earth having wasted away, and only the bare framework of the land being left.'

The Romans were less environmentally conservative than the Greeks and showed scant concern over the potential deleterious consequences of their activities. They, like the Christians who followed them, took a possessive view of Earth — it was the property of *Homo sapiens*, to be exploited in any way that people wished. During the Roman Empire, deforestation spread from the hills of Galilee and the Taurus Mountains of Turkey in the east to the Sierra Nevada of Spain, under the impetus of the Roman agricultural system.

Roman society had little compensating ethic of conservation, because beauty and utility were tied together in much of Roman thought, and Romans took great pride in creating a 'second nature' through their own efforts. Pliny the Elder noted that human beings sometimes abused Earth, their mother, but he and most Romans saw the abuse simply as a failure of intelligent husbandry. This attitude has survived and still dominates Western thinking about land use and management.

The Egyptians and Greeks were determined hunters of big game and were responsible for local extinctions of some large animals, such as the lions of Greece and upper Egypt. But the Romans far outdid their predecessors in hunting for meat, skins, feathers and ivory. Even more important may have been the multitudes of beasts captured for use in 'games'. The scale of these brutal entertainments, which pitted animals against one another and against people, is hard to grasp from a distance of two millennia. Titus dedicated the Colosseum with a three-month series of events in which 9,000 beasts were killed. And the celebration of Trajan's conquest of Dacia (modern Romania) involved games in which 11,000 wild animals were slaughtered.

These numbers only suggest the scope of the destruction, however. The conditions of hunting, transport and housing of animals in Roman days must have meant that for every lion, tiger, leopard, rhino, buffalo, giraffe or hippo that died entertaining the masses, dozens or even hundreds of others must have perished before reaching the arena. The Roman Empire was probably responsible for the greatest die-off of large animals since the Pleistocene extinctions. While there is no evidence that any species of large mammal was wiped out by the Empire, numerous populations were destroyed or decimated, and the ranges of many species were therefore severely constricted.

Just as the Mesopotamians paid a high price for their inability to

159

maintain their life support systems, so the Romans suffered for their short-sighted exploitation of the environment. Erosion and exhaustion of soils as well as deforestation were all factors in the decline and fall of the Roman Empire. But at least the Greeks and Romans had some appreciation of such things as the role of forests in maintaining watersheds; they in fact planted the first seeds from which the biological sciences, including ecology and modern natural history, sprang.

In contrast, the Europeans of the Middle Ages paid little attention to such matters and pushed ahead with the destruction that the Romans had done so much to promulgate. The Romans left about a third of Britain forested, for instance. The Angles and Saxons took it from there, and today less than a tenth is forested. The use of large trees for the masts of naval and merchant ships in the late Middle Ages helped speed the destruction of forests all over Europe.

Five centuries ago, when the Europeans had overpopulated their continent and were pushing their own resources towards exhaustion, they began to spread over Earth in a campaign of economic exploitation of frontiers that continues with only slight modifications today. The arrival of Europeans in North America was quickly perceived by the American Indians not just in terms of conquest, brutality and enslavement, but as a threat to their very way of life.

Like all hunting and gathering peoples, Native Americans had a keen eye for the state of their environment. Many plains Indians, for example, understood that killing off the bison herds would make survival impossible for themselves, even if they should be so fortunate as to win their battles against the advancing white men. The long covered-wagon trains crawling along the Oregon trail were accompanied by massacres of game, and the settlers continually abused the natives, who were well aware from contacts with eastern tribes of the fate of Indians where the Europeans dominated.

In their hour of need, the small Native American populations produced an astonishing array of brilliant leaders, men such as Hiawatha, King Philip, Popé, Pontiac, Tecumseh, Osceola, Black Hawk, Chief Joseph, Crazy Horse and Geronimo. They resisted fiercely but hopelessly – their last great victory being that of the Cheyennes and Sioux under Crazy Horse over that famous Lieutenant Colonel of the 7th Cavalry, George Armstrong Custer. Custer was, one might claim, a suitable representative of Western Civilisation – arrogant, seemingly without introspection, and determined to attack regardless of the consequences.

But the Indian cause was hopeless, and the victory on 25 June 1876 on the Little Big Horn river was the beginning of the end. Fifteen years later, both the free-roaming Indians and the free-roaming plains buffalo were no more, and the destruction of the North American environment has accelerated ever since.

We have no written record of Crazy Horse's view of the Europeans; his deeds must speak for him, because he was murdered at the age of 35 by people who could never conquer him. But one might imagine that great Lakota Sioux leader would agree with Chief Sealth of the Duwamish Tribe of Washington State, who did leave a written opinion. Sealth, whose name lives on in that of the city of Seattle, wrote in a letter to President Franklin Pierce in 1855:

> Every part of the earth is sacred to my people. Every shining pine needle, every sandy shore, every mist in the dark woods, every clearing and humming insect is holy in the memory and experience of my people. The white man . . . is a stranger who comes in the night and takes from the land whatever he needs. The earth is not his brother but his enemy, and when he has conquered it, he moves on. He leaves his father's graves, and his children's birthright is forgotten . . . all Things share the same breath — the beasts, the trees, the man. The white man does not seem to notice the air he breathes. Like a man dying for many days, he is numb to the stench. . . . What is man without the beasts? If all the beasts were gone, men would die from great loneliness of spirit, for whatever happens to the beast also happens to man. All things are connected. Whatever befalls the earth befalls the sons of earth. . . . The whites too shall pass — perhaps sooner than other tribes. Continue to contaminate your bed, and you will one night suffocate in your own waste. When the buffalo are all slaughtered, the wild horses all tamed, the secret corners of the forest heavy with the scent of many men, and the view of the ripe hills blotted by talking wires, where is the thicket? Gone. Where is the eagle? Gone. And what is it to say good-by to the swift pony and the hunt, the end of living and the beginning of survival.

In spite of the exploitative 'there's always a new frontier' attitude developed by Western societies, some members of those societies had begun to appreciate some environmentally negative aspects of the Western notion of 'progress' even before the time of Chief Sealth. Scattered attempts to preserve animals go back centuries before Christ to the time of the Assyrians — but they appear to have been attempts primarily to provide rich hunting for the aristocracy. In 1534, Parliament under Henry VIII passed the first conservation legislation in the West — an act to protect the eggs of wild birds, in part to serve the interest of nobles in falconry. Sporadic attempts were made in various places, including Britain, to preserve dwindling forests, but not until the nineteenth century did widespread concern for diverse aspects of the environment begin to take shape.

In Britain, a combination of an intense interest in natural history

99. **Chief Sealth.** In a letter that in some ways anticipated the modern ecological movement, this Indian chief rebuked the President of the United States for the exploitative behaviour and attitudes of white men.

100. **Charles Darwin.** Darwin's coherent, heavily documented ideas on evolution inspired an interest in nature and an appreciation for it among many of his contemporaries in Western societies.

101. **The tiger hunt.** People in the Victorian era generally regarded wild animals as curiosities or as something to be conquered or exploited. This 1887 picture, depicting the thrill of the hunt (and thereby perhaps enhancing the perceived value of a tiger skin), was an advertisement for the International Fur Store.

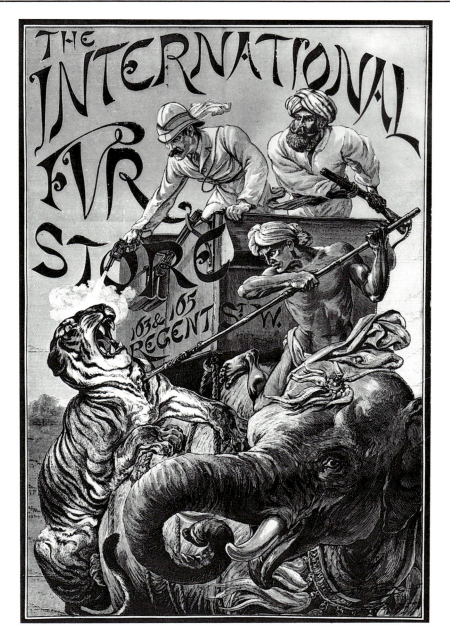

and a growing desire to protect domestic animals from cruelty coalesced into a general campaign for the conservation of wildlife. The specimen-collecting urges of Victorian naturalists provided evidence of the decline of wildlife and spurred the formation of societies to preserve it. Perhaps the first in the world was the East Riding Association for the Protection of Sea Birds, formed in 1868 in an attempt to stop the annual slaughter by 'sportsmen' of nesting kittiwakes, guillemots and gulls.

Other organisations followed quickly, and by 1891 several had fused into the powerful (and since 1904, Royal) Society for the Protection of Birds. By the turn of the century, the widespread notion that nature existed in a robust 'balance' was, at least in Britain, well on its way to being replaced by a new one focussed on the fragility of wild species. In 1915, the distinguished zoologist Sir Ray Lankester expressed this new view of Earth: 'Once man is present in the neighbourhood, even at long distance, he upsets the "balance of nature".'

During the late 1800s, the impact of humanity on Earth was probably most comprehensively delineated not by Englishmen, but by an American, George Perkins Marsh. His great work, *Man and Nature; or, Physical Geography as Modified by Human Action* was published in 1864, and its revised edition, *The Earth as Modified by Human Action*, appeared a decade later. The preface to his first edition pretty much says it all. Marsh's objectives could be the ones for this book:

> . . . to indicate the character and, approximately, the extent of the changes produced by human action in the physical conditions of the globe we inhabit; to point out the dangers of imprudence and the necessity of caution in all operations which, on a large scale, interfere with the spontaneous arrangements of the organic and inorganic world; to suggest the possibility and the importance of the restoration of disturbed harmonies and the material improvement of waste and exhausted regions; and, incidentally, to illustrate the doctrine that man is, in both kind and degree, a power of a higher order than any of the other forms of animated life, which like him, are nourished at the table of bounteous nature.

Two of his themes are very familiar – human impact on other species and especially on forests. He opens his work with the comment: 'The action of man upon the organic world tends to derange its original balance, and while it reduces the numbers of some species, or even extirpates them altogether, it multiplies other forms of animal and vegetable life', and 'felling of the woods has been attended with momentous consequences to the drainage of the soil, to the external configuration of its surface, and probably, also, to local climate'.

Interestingly, the debate about governmental versus private responsibility was as alive in Marsh's mind as it is today within the environmental movement. He opined that:

> Joint stock companies have no souls; their managers, in general, no consciences. . . . In fact every person conversant with the history of these enterprises knows that in their public statements falsehood is the rule, truth the exception.

Marsh's vote clearly went to government:

102. **Nathaniel Charles Rothschild.** Charles Rothschild was a leading naturalist who helped found a society for the Promotion of Nature Reserves in Britain and the Empire in 1912, essentially Britain's first serious effort to conserve ecosystems rather than individual species, and one that arose mainly from Rothschild's vision. He was also instrumental in forming the international association that eventually became the International Union for the Conservation of Nature (IUCN). Although he spent most of his life in the family's banking business, Charles Rothschild found time to maintain his natural history studies and write over 150 scientific papers. Unfortunately, owing in part to Rothschild's untimely death in 1923, Britain did not join the precursor of the IUCN, and conservation efforts within the country did not move very far until decades later. Rothschild's daughter, Miriam, is today a distinguished evolutionary biologist and a Fellow of the Royal Society.

No doubt . . . organisation and management . . . by government are liable, as are all things human, to great abuses. The multiplication of public placeholders which they imply is a serious evil. But the corruption thus engendered, foul as it is, does not strike so deep as the rottenness of public corporations.

One cannot help but wonder what Marsh would say today. Our guess is that he would find both governments and corporations more of a mixed bag.

103. **Father Thames.** As this 1855 cartoon from *Punch* shows, pollution was a problem in Victorian England, and people were very much aware of it.

FARADAY GIVING HIS CARD TO FATHER THAMES;
AND WE HOPE THE DIRTY FELLOW WILL CONSULT THE LEARNED PROFESSOR.

Marsh was probably the most far-sighted environmentalist of his century, in many ways the first modern environmentalist. His perceptions are all the more impressive considering that, when he wrote, the United States still had substantial unexploited frontier areas. A century and a quarter ago, Marsh basically had it right: *Homo sapiens* was supported by Earth's ecosystems but had the power to destroy them. He saw clearly that humanity would have to pay close attention to what it did to Earth or it would pay the consequences. The passage of time has only confirmed the accuracy of his message and made more relevant than ever his observation that 'the multiplying population and the impoverished resources of the globe demand new triumphs of mind over matter'.

Coincidentally, one year after the publication of *Man and Nature*, the Yosemite Valley of California was preserved by Congress for public recreation. And in 1872, before the second edition of Marsh's classic book appeared, Yellowstone was declared the first national park. In those days, tension was rising between 'preservationists' and 'conservationists' in America. Preservationists were typified by John Muir, defender of Yosemite and one of the founders in 1892 of the Sierra Club, whose members share the tradition. They were dedicated to protecting Earth's riches for their own sake and for the psychic benefits people could gain from wilderness. Conservationists, by contrast, led by the great forester Gifford Pinchot, focussed on the wise development of natural resources.

Conservation eventually became the official policy of the United States, while preservation remained, until rather recently, in eclipse. The victory of conservation was seen in the damming of the beautiful Hetch Hetchy valley in Yosemite National Park after a long and highly publicised battle, starting in 1908, over what was the best use for the valley – supplying a dependable drinking water supply for San Francisco or preserving one of the great beauty spots of Earth. The opposing views were well expressed by the champions of preservation and conservation:

Muir: These Temple destroyers, devotees of ravaging
         commercialism, seem to have a perfect contempt for Nature,
         and instead of lifting their eyes to the God of the mountains,
         lift them to the almighty dollar.
Pinchot: The injury . . . by substituting a lake for the present swampy
         floor of the valley . . . is altogether unimportant compared
         with the benefits to be derived from its use as a reservoir.

After a bitter national debate, the conservationists won, the dam was built, and (as the preservationists had predicted and the conservationists denied) the beauty of the valley was largely destroyed.

The debate was historic for two reasons. First, that it occurred at all in a nation with a barely closed frontier, to say nothing of the magnitude of the uproar, was a measure of the grip that the idea of 'wilderness', as a positive value, already held on the American mind. Second, it demonstrated clearly that, environmentally, as in other areas of human endeavour, one often cannot have one's cake and eat it too. Most of those involved in the Hetch Hetchy debate recognised that two

165

'goods' were involved; the dispute was over the priorities to be assigned to two potential uses of Hetch Hetchy.

Even after more than half a century, that question remains unresolved. Ideally, California's population growth and development (and those of the nation and world with which that great state interacts) should have been planned to avoid the crowding and shortages of clean air and fresh water that now plague the state. It might also then have avoided desecration of some magnificent wilderness. But in a less-than-ideal world, the flooding of the Hetch Hetchy valley was probably inevitable, and from a human viewpoint, 'right'. Even today, few people are far-sighted enough to contemplate the necessary population reduction and reorganisation of California that would be required for that state to be relatively secure against environmental disaster.

In the early twentieth century, the discipline of ecology began to develop the scientific background for the environmental movement. Plant ecologists led the way on both continents. Men like F.E. Clements in the United States and A.G. Tansley in Britain studied the characteristics of ecological systems and the successional changes that occurred in them. It was in Britain in the 1920s, however, that a young man who was to become the greatest ecologist of the mid-century, Charles Elton, began to tie the ecology of other organisms to human ecology.

The seeds were sown in Elton's 1927 classic, *Animal Ecology*, the first readable text on the subject. Thirty years later, *The Ecology of Invasions by Animals and Plants* completed the connection. Of much broader significance than its title might imply to a layperson, it closed with a chapter on 'The Conservation of Variety'. At the end, Elton the scientist refers to the leader of preservationists:

> Would it not be good to be able to say like John Muir, the Scotsman who became the great American prophet of wilderness conservation: 'To the sane and free it will hardly seem necessary to cross the continent in search of wild beauty, however easy the way, for they find it in abundance wherever they chance to be.' Will we be able to talk like this in fifty years' time, as he could fifty years ago?

Happily, Elton, a shy and retiring man, lived long enough to see ecology become one of the cutting edges of the biological sciences and environmentalism a great popular movement.

Perhaps the most important American contemporary of Elton was the brilliant, eloquent, applied ecologist, Aldo Leopold. He might be roughly described as a combination wildlife manager and George Perkins Marsh who understood the developing field of ecology. Indeed, in 1931 he met Elton at a conference on biological cycles and used

Elton's ideas in attacking practical problems. Leopold was the first 'populariser' who saw that *Homo sapiens* was both part of natural ecosystems and dependent upon them. He pleaded that ethical considerations be extended by people and societies to the natural environment – to Earth itself. Leopold's 'land ethic', as he explained,

> . . . simply enlarges the boundaries of the community to include soils, waters, plants and animals, or collectively: the land. . . . In short [it] changes the role of *Homo sapiens* from conqueror of the land community to plain member and citizen of it. It implies respect for his fellow-members, and also respect for the community as such.

Aldo Leopold also penned words that ring a bell with all people who are ecologically knowledgeable but find themselves debating endlessly with others who believe that humanity can continue forever on its present growth-manic course.

> One of the penalties of an ecological education is that one lives alone in a world of wounds. Much of the damage inflicted on land is quite invisible to laymen. An ecologist must either harden his shell and make believe that the consequences of science are none of his business, or he must be the doctor who sees the marks of death in a community that believes itself well and does not want to be told otherwise.

Amen.

The postwar Western view of the human predicament, especially its international dimensions, can be traced in part to two other American writers of the mid twentieth century – William Vogt and Fairfield Osborn. Their books, *The Road to Survival* and *Our Plundered Planet* (respectively), first introduced many people to the overall issues of population growth, the depletion of non-renewable resources, and the human assault on Earth's life support systems. The end of *Our Plundered Planet* is worth contemplating today, even though it was published in 1948:

> Technologists may outdo themselves in the creation of artificial substitutes for natural subsistence, and new areas, such as those in tropical or subtropical regions, may be adapted to human use, but even such recourses or developments cannot be expected to offset the present terrific attack upon the natural life-giving elements of the earth. There is only one solution: Man must recognise the necessity of cooperating with nature. He must temper his demands and use and conserve the natural living resources of this earth in a manner that alone can provide continuation of his civilisation. The final answer is to be found only through comprehension of the enduring processes of nature. The time for defiance is at an end.

While the works of Leopold, Vogt, Osborn and those who went before attracted relatively little public attention, they greatly influenced our

167

own thinking and that of many other ecologists of our generation.

Soon after, the first major environmental issues of the post-war world shifted from the pages of books to the headlines of newspapers. Air pollution, in the form of a smog disaster in Donora, Pennsylvania in 1948, killed twenty people outright and injured many more. Then the London smog disaster of 1952 caused some 4,000 deaths, and the growing problem of smog in Los Angeles (at first treated as a joke) gradually became a public issue.

104. **London's killer fog.** In December 1952, some 4,000 excess deaths in the London area were attributed to the mix of toxic pollutants accumulated in a heavy fog that blanketed the city for many days. The poisonous fog was so thick that policemen had to guide bus drivers through the streets. The event increased concern over environmental hazards throughout the developed world.

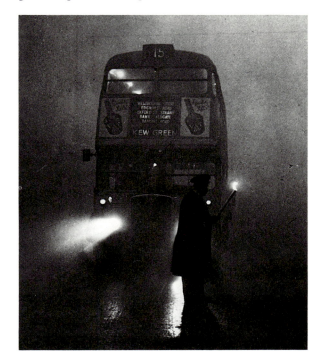

About the same time, radioactive fallout from nuclear weapons testing became an important matter of public debate. Testing of nuclear weapons in Nevada in the early 1950s produced radioactive rainstorms over some American cities, and monitoring soon showed 'hot' fallout over much of the nation. Over the next decade, scientists, including biologist Barry Commoner, who later became one of the first 'visible scientists' of the environmental arena, led a battle to confine nuclear testing underground. Their crusade was largely successful; in 1963 the Limited Nuclear Test Ban Treaty ended the exploding of nuclear weapons in the atmosphere by the United States, the United Kingdom and the Soviet Union (France and China, non-signatories of the treaty, continued atmospheric testing).

However, it took the widespread, blatant misuse of pesticides to move ecology firmly and permanently onto the political agenda of the United States. In the late 1940s and early 1950s, DDT was viewed as the ultimate panacea for dealing with humanity's insect enemies. We can well recall going to drive-in movies in Kansas during that period and having a cloud of DDT blasted through the open window of our car by a gigantic portable blower. The possible deleterious effects of DDT and other synthetic poisons on human beings were discounted, and the warnings of ecologists and evolutionists about the problems that would plague insect-control programmes if the misuse of pesticides continued, were ignored.

A typical case was the massive and supremely foolish attempt by the United States Department of Agriculture (USDA) to eradicate the fire ant from the south-eastern United States in 1957. The ant was and is a nasty but not-too-serious pest in the south-eastern United States. Its mounded nests make cultivation of farm fields difficult, and some sensitive people have become severely ill or even died as a result of their stings (although fire ants are less of a menace in this regard than bees and wasps).

The USDA developed the astonishing idea that, by aerial spraying of dieldrin (a potent relative of DDT) over much of the south-eastern United States, the fire ant could be eradicated. Led by Edward O. Wilson, a specialist in the biology of ants, ecologists protested against the plan. We wrote a long letter to the Secretary of Agriculture saying, among other things, 'the fire ant is one of the least likely of the insects in the area to be exterminated'. The biologists were ignored, the programme was carried out, and the predicted results were obtained: desirable wildlife was decimated in many areas, and the fire ant thrived and spread.

The USDA and state agriculture departments have continued to make bungled assaults on the fire ant. The main result has been to waste about $150 million of the taxpayers' dollars. Ed Wilson, who has since become world-famous as the founder of the discipline of sociobiology and as an outspoken defender of Earth's diversity of organisms, later described the fire ant control programme in the south as the 'Vietnam of entomology'.

In 1962, shortly after the USDA began to get bogged down in its 'Vietnam', the spark came that ignited the modern environmental movement. Rachel Carson published her beautifully written critique of pesticide use, *Silent Spring*. The time was ripe, and Carson's eloquent plea for a re-evaluation of policies of pesticide use hit home by accurately casting the public's most beloved group of wild animals — songbirds — as the first prominent victims of a rapidly building disaster.

The decade following *Silent Spring* was the decade of a great environmental awakening. It contained many shocks beyond the

169

105. **Rachel Carson.** With her brilliant book, *Silent Spring*, biologist Rachel Carson awakened the American public to the dangers of broadcast spraying of pesticides and launched the environmental movement of the 1960s. She was mercilessly abused by the pesticide industry, but her basic arguments have stood the test of time.

106. **Detergent foam in the streets.** Occurrences such as detergent foam blown from the river into the streets of Castleford, Yorkshire, inspired public outcries about environmental deterioration in the late 1960s and early 1970s in many Western countries. In this instance, a change in the formula of the detergent in the mid-1970s removed the problem.

170

realisation, fostered by Rachel Carson, that environmental deterioration meant more than foul air and water; it could threaten one's very life and livelihood.

In 1966, a coal slag heap slumped onto the village of Aberfan, Wales, claiming 144 human lives. A year later, almost a million barrels of crude oil were spilled onto the western approaches of the English Channel when a Liberian tanker ran aground off Cornwall. According to distinguished British conservationist Max Nicholson, the disaster caused 'a financial waste of around £10 million and inflicted perhaps the greatest single injury ever caused to the tourist and holiday industry in peacetime, apart from incalculable damage to the natural environment and to wildlife'.

America's turn came two years later when a blowout at an undersea well injected a few thousand tons of oil into the beautiful Santa Barbara channel. The location off a wealthy, resort-oriented community had an impact far beyond the scale of the accident; it underlined the need for laws to protect the environment. Perhaps even more dramatic was the discovery by Thor Heyerdahl, as he sailed a papyrus raft across the Atlantic in 1970, that 'clots of oil are polluting the mid-stream current of the Atlantic Ocean from horizon to horizon'.

Recognition of environmental problems was hardly confined during this period to the United States and Great Britain. Western Europe was rocked by a huge fish-kill in the Rhine river caused by pollution. Japan reeled under revelations of serious smog incidents, the Minimata mercury disaster and poisonings by PCBs, cadmium and other substances.

The mounting awareness of environmental problems in the late 1960s and early 1970s ignited a mass movement in the United States. On the first Earth Day, 22 April 1970, millions of people participated in over 10,000 events. David Brower, the charismatic leader who founded 'Friends of the Earth' in 1969 after the Sierra Club removed him as president in protest over his radical actions, organised volunteers and professional staff to press for environmental legislation. The Sierra Club and its fellow conservation organisations followed suit. Meanwhile, some public-interest organisations, such as the Environmental Defense Fund and the Natural Resources Defense Council, began winning battles against irresponsible corporations and government agencies through litigation in the courts.

During the same few years, people had begun to focus on the population dimension of environmental issues – and the role of population growth in exacerbating the plight of poor countries. The publication in 1968 of Paul Ehrlich's book *The Population Bomb* brought home to many people the critical role that the runaway increase in human numbers was playing in causing environmental deterioration

171

107. **Alaskan oil pipeline.** The proposal to build an oil pipeline several thousand miles long to bring oil from the Alaskan North Slope to the port of Valdez in southern Alaska in the late 1960s led to an intensive environmental impact review of the project through the courts. Environmentalists imposed many changes in the project to mitigate environmental damage and protect wildlife, changes that the builders and operators later admitted had resulted in economic benefits to them. Here the Alyeska Oil pipeline runs down towards the Tazlina River.

and rapid resource depletion. Shortly thereafter, the public-interest organisation 'Zero Population Growth' was founded, with the specific goal of bringing the American birth rate into balance with the death rate.

Planned Parenthood organisations had long been active in many nations, of course, making available to individual couples the means and information to limit their families. The concept of limiting the growth of a population, however, was new in the 1960s, and was resisted in many quarters. A report in 1972 by the national Commission on Population Growth and the American Future concluded that there was no compelling reason for further growth in the American population and that continued growth was likely to intensify many already existing problems. And this conclusion, remarkable for its time, was reached after only a very superficial exploration of environmental and resource constraints on growth.

The climax of the era of ecological awakening is perhaps best marked by the publication of the Club of Rome's famous study, *The Limits to Growth*, by Dennis and Donnella Meadows of Dartmouth University, New Hampshire, in 1972. Computers, then rather novel and revered devices, came to the same obvious conclusions as had ecologists – that the size of the human enterprise could not be infinite in a finite universe. That work, reinforced by the lessons of the oil crisis of 1974, firmly established population-resource-environment issues in the American consciousness. That consciousness soon spread to the rest of the world.

In general, the 1970s were a period of rapid advance in attempting to deal with these issues. Western nations were beginning to grapple with environmental problems by enacting serious anti-pollution regulations. Internationally, concern found expression in the first United Nations Conference on the Human Environment, held in Stockholm, Sweden in 1972. While the conference was a scene of disharmony, in which delegates from developing countries maintained that environmental deterioration was no concern of theirs, it did raise global consciousness to a large degree and gave birth to a new agency, the United Nations Environment Programme (UNEP).

In the United States, in response to the public's rising concern, a quite remarkable series of environmental laws was enacted by the United States Congress. The first, in 1969, was the National Environmental Protection Act (NEPA), which among other precedents established the responsibility of the federal government for maintaining a clean and healthy environment. The requirement of an 'environmental impact statement' for any prospective government-built or -funded project was part of NEPA. In addition, the Environmental Protection Agency (EPA), the nation's environmental regulatory agency was established, as was the Council on Environmental Quality, attached to the White House and charged with monitoring the nation's environmental quality and reporting to the president and the people.

Soon after NEPA came new laws for clean air and clean water, a tightening of regulations on pesticides, the banning of DDT and several other chlorinated hydrocarbon pesticides, and the Endangered Species Act. The last was a particularly important step forward because, not only did it extend protection to threatened and endangered species, it emphasised the preservation and, if necessary, restoration of habitats. Moreover, through a clause to protect subspecies, it offered protection for rare populations of species that were not endangered in all parts of their ranges. In addition, a series of laws was passed, setting aside wilderness areas and extending special protection to the sea coasts and fragile barrier islands.

In the late 1970s, Congress addressed the latest of the serious environmental threats to emerge – the problems of toxic substances and wastes – by passing two new laws, the Resource Conservation and Recovery Act (RCRA), which emphasises recycling and proper disposal, and the Toxic Substances Control Act (known as TSCA, or often TOSCA), providing for registration, assessment of hazard and monitoring of all chemicals manufactured and marketed in the United States. The Occupational Safety and Health Act (OSHA) regulates hazards in workplaces, and a law with an impossible acronym, but better known as the 'Superfund', was passed to fund the clean-up of chemical waste dumps, of which an estimated 50,000 exist in the United States.

173

108. **Toxic waste disposal.**
Leaking containers of toxic
wastes from an
inadequately designed
dump are being relocated
by workers for the US
Environmental Protection
Agency. A conservatively
estimated 50,000 dump
sites in the United States
need to be assessed and
improved. The Reagan
Administration has done
very little to deal with the
problem, despite the
existence of laws requiring
it. The British Government
has yet to come to grips
with the problem at all.

109. **Greenpeace in action.**
Perhaps the most militant
of environmental
organisations, members of
the international group
Greenpeace are here seen
harassing an Icelandic
whaling vessel in 1978.
Pressure from Greenpeace
helped to achieve an
international agreement for
an end to all commercial
whaling by 1986. The
photographer who took this
picture, Fernando Pereira,
was tragically killed when
Greenpeace's anti-nuclear
flagship, *The Rainbow
Warrior*, was bombed and
sunk in Auckland harbour,
New Zealand, by two
French agents in 1985.

The problems of environmental deterioration in the United States have not by any means been solved, although legal tools have been developed that have allowed many aspects to be improved. But putting laws in place is only the first step, as Americans learned in the 1980s; after that come implementation and enforcement, which often entail a series of hard choices.

A perpetual problem in environmental policy, one yet to be entirely resolved by any society, is the dilemma of inequity. The benefits and costs of both destructive and constructive actions often fall on quite different segments of the population. One segment may be quite clearly defined, such as particular industries that either make more money by pushing some of their costs onto the general public (by not controlling their effluents) or are forced by that public to bear unexpected costs (because effluents once thought benign turn out to be dangerous). The other segment is often the 'general public'. While some members of the public buy the goods produced by the industry, others may be forced to pay the price for the pollution. An obvious

110. **A Friends of the Earth demonstration.** Members of the UK group of this international activist environmental organisation deposited a ton of bones on the doorstep of the Department of the Environment in 1981 to protest about the Government's laxness in protecting endangered species.

175

way to make action more palatable and fair is for the public to help shoulder the costs of environmentally sound actions forced on individuals, corporations or political entities, if they previously behaved responsibly.

For example, when the industrial Midlands of England, the United States Midwest and the Ruhr district of Germany were industrialised, it was considered perfectly proper for factories to discharge a rich mix of pollutants from their smokestacks. As time passed and the public demanded cleaner air, the companies complied – often by building taller smokestacks. Those stacks greatly ameliorated the perceptible air pollution near the plants but they actually made the problem of acid precipitation worse. Now further steps are required, but heavy industry in many nations is in economic trouble, and the costs of abating acid rain in some cases may threaten a company's ability to survive.

A similar situation has arisen with toxic wastes. Decades ago, such wastes were commonly incinerated. When the public demanded control of the air pollutants produced by incineration, 'sanitary' land fills were substituted as repositories of wastes. Unfortunately, many of the land fill operations were inadequately designed or inappropriately located, and toxic wastes oozed out, often onto neighbouring property and into local aquifers. As with acid rain, the solution proved worse than the original problem, and even costlier to cure.

The answer to this sort of situation is for each nation as a whole to make a substantial contribution to the burden of pollution control, since the nation as a whole will benefit from it. But the process of persuading the public that the cost is justified in a particular case may be a long one, especially when the issue is as complex and subtle as the effects of acid precipitation, despite the clear evidence of public support for environmental protection in general. The acid rain issue is further complicated by being an international one; abatement in one nation must, to be both fair and effective, be matched by abatement in its neighbours.

Since 1970, Britain, like the United States, Japan and most European nations, has imposed increasingly tight controls on effluents from factories and vehicles and the use of toxic materials such as pesticides. The Department of the Environment was formed in 1970, but its activities have mainly focussed on urban environmental management. A plethora of government and quasi-governent agencies has been established for managing and protecting lands in a variety of categories, including national parks, recreation areas, nature reserves, forests, scenic areas and areas of particular scientific interest, abetted by laws such as the Wildlife and Countryside Act. In aggregate, these areas amount to more than 40 per cent of the nation's land, but for most the level of protection leaves much to be desired.

While the environmental movement has been almost universal in the West, its expression has been quite different in various nations. In some, particularly those with parliamentary systems of government featuring several political parties, action has centred around new political parties whose principal concerns are environmental – the so-called 'Green' parties. These movements have been strongest in France and West Germany, where Green parties have won 5 per cent or more of the vote and on occasion have dominated in local elections. Green parties have also had some success in Australia and New Zealand. Besides embracing the traditional environmental goals, many of the Green movements, especially in West Germany, also oppose nuclear power and are advocates of nuclear arms control.

111. **The Greens parade.** The Green party in West Germany has been gaining in strength since the 1970s. Here they participate in a giant peace march in Bonn in 1981. The poster reads: 'Ecology means peace – for a nuclear-free Europe – the Greens'.

Environmental protection laws were also enacted in Eastern European nations and the Soviet Union during the 1970s, but they do not seem to be strictly observed. The communist governmental structure produced a classic case of the fox guarding the henhouse, with the same entity (the government) functioning as both perpetrator (developer) and policeman (environmental protection enforcer). The Soviet Union's showplace, Lake Baikal, which contains the world's largest volume of fresh water and a unique aquatic fauna, was threatened by emplacement on its shore of a polluting paper and pulp mill with little or no control of its effluents. The mill was built, but the spotlight of international attention and the protests of the Soviet

177

scientific community apparently led to the installation of pollution controls. For the most part, the Soviet government and the governments of its satellites do as they please in environmental protection; the people are poorly informed of problems and have little recourse when they do learn of abuses.

By the 1980s, attitudes in the developing nations towards environmental concerns had changed dramatically. Much of this change was due to the steadily declining per capita food production and recurrent drought and famines in sub-Saharan Africa, culminating in the disastrous, continent-wide famines of 1984–5, as recognition grew of the consequences of environmental neglect and land abuse. These disasters also at last awakened African leaders to the dangers of overpopulation, which they had previously complacently considered to be a problem for the distant future, if ever.

The need to preserve some portion of the rapidly dwindling natural areas in the tropics – especially tropical forests – also began to be recognised by some developing nations in the late 1970s and early 1980s. A leader in this was Costa Rica, which set aside some 10 per cent of its land area in nature reserves. Brazil, too, which until 1980 had vociferously asserted its right to plunder its resources in its own way, began to change its policy. But such a complete reversal in treatment of the environment is not easily made in a country with a rapidly increasing population, much of it desperately poor, and a strong industrial element of society whose firmly entrenched attitude is that resources exist to be exploited.

International efforts to control pollutants and institute environmental protection have continued since the first United Nations Conference on the Human Environment in 1972, including a second conference ten years later. UNEP has operated with quiet diplomacy in working out important regional environmental agreements in several parts of the world, notably the Mediterranean Basin. That compact entailed the cooperation of several nations that have no diplomatic relations with each other. Other significant steps include agreements to control the international transport of endangered species or their products (such as elephant tusks, rhinoceros horns or the pelts of large cats).

In addition to these actions by official international groups, private organisations such as the World Wildlife Fund have played key roles in bringing about change. The latter group has long been active in gaining international cooperation to protect endangered and threatened species. More recently, it has shifted its focus to emphasise the preservation of 'habitat' – natural ecosystems that so far have survived the human onslaught.

In 1980, the International Union for the Conservation of Nature and Natural Resources (IUCN) and the World Wildlife Fund collaborated in

112. **Nature on television.** Television productions on nature and wildlife are among the most popular programmes, especially with children. This popularity has risen in tandem with increasing environmental consciousness in developed countries.

drawing up a World Conservation Strategy, a proposal for nations to plan their development in ways that would preserve biotic diversity and protect natural resources. So far, unfortunately, only a few nations have accepted the challenge to draw up and implement national conservation strategies patterned on it, and those few, including the United Kingdom and Australia, have produced only draft proposals.

A great deal of progress has clearly been made since 1970 in raising awareness of environmental problems and in establishing a clean and safe environment as a basic human right. Further, in many countries, environmental laws are in place and being enforced at some level. In less than two decades, environmental protection advanced from being a semi-revolutionary concept to an almost fully institutionalised part of

113. **A protected tiger.** This tiger, one of an extremely endangered species, was photographed at Tiger Tops Lodge in Royal Chitwan National Park, Nepal. Visitors to the park can see the animals in the preserve from the back of an elephant, as tiger hunters once did.

governing a modern state. For a while, it appeared that civilisation might move fast enough to preserve Earth as a habitable environment for *Homo sapiens*. But the forces of reaction were stronger than they appeared in the early 1970s, and they have put Earth back on the road toward ecological catastrophe.

CHAPTER EIGHT

The Sprint of Folly

We have seen that the environmental movement has deep roots, and that the situation described in the early parts of this book is unfolding just as many observers have warned that it would. But, rather than face up to the human predicament, many people still prefer either to deny its existence or to claim that, although it exists, it can easily be solved by technological miracles that are just around the corner. After all, hasn't Western civilisation already provided a previously unheard-of level of affluence for hundreds of millions of people? Surely nothing is beyond its capabilities!

The notion that everything must be just fine is very deeply embedded both in the ethos of our civilisation and in the minds of those whom the socio-economic system of that civilisation has bubbled to the top. The rich and powerful quite naturally view their own positions as proof-positive that the system must be working well. And the poor have little time to think about the system at all – they are much too busy trying to survive.

A tendency to ignore unpleasant trends seems ever present in human societies. Furthermore, today's human beings have neither been designed by evolution nor trained by their culture to deal with long-term trends. It is difficult for most people to view the plight of humanity in a historical, to say nothing of an evolutionary, perspective.

Barbara Tuchman, in her marvellous book *The March of Folly*, defined a political course of action as 'folly' if it had three characteristics: first, it was recognised as a mistake in its own time; second, a reasonable alternative course of action was available; and third, the policy was not that of a single ruler and would outlast a single political lifetime. All of the actions and attitudes discussed in this chapter fulfil those requirements, but there is no question that the current march of folly has turned into a sprint in the present decade.

Why has folly so easily become a way of life? The increase of ecological awareness over the past few decades, first in Western nations and later elsewhere, has come during a period of unprecedented economic expansion. The two, of course, have been closely related. The post-World War II economic boom stimulated population growth in the rich nations, while death control technologies such as antibiotics and DDT, first deployed during the war, fuelled the population explosion in poor nations. Many factors, including escalating numbers of private cars on the roads and increasing consumption in general began to spread and exacerbate environmental problems that had their roots in the industrial revolution.

114. Military waste. These F106 aeroplanes are in 'mothballs' at the Aerospace Maintenance and Reutilization Center, Davis Monthan Air Base, Arizona. The Reagan administration believes that piling up arms is what makes the United States strong, regardless of the utility of the weapons – and regardless of the condition of the natural systems and resources that are the real source of the nation's strength.

But perhaps more importantly, increased leisure, rising levels of education and a flowering of medicine and science, all related to economic prosperity, made people in the overdeveloped countries more aware of the deleterious side-effects of economic activity. The benefits flowing from the post-World War II economic boom were more widely shared than those of any previous spurt of growth. They transformed the lives of many citizens in the most powerful nations of the world, bringing wealth and power to a minority of human beings, but a minority that for the first time was large enough to make up a substantial proportion of Earth's population.

Naturally, the benefits of economic growth have been enjoyed by those who received them and desired by those who did not. The benefits are easily perceived: manufactured goods such as cars, refrigerators, radios and television sets; access to education, good medical care, high-speed transport and so on. The costs, in contrast, have not always been so apparent.

Some of the more obvious costs, such as air and water pollution, are often paid by less influential people who also receive fewer of the benefits. It is the poor who are usually forced to live near factories and downwind of power plants. It is the poor who cannot afford to fly to distant beaches for vacations, but are allowed to swim in polluted water near industrial centres. It is the poor whose choices have been

115. **An early assembly line.** Cheap power from fossil fuels, combined with the efficiency of assembly-line production, made possible personal transport in the form of private cars in the industrialised nations, especially the United States, before the middle of the twentieth century. This resource-intensive possibility eventually became almost obligatory as governments subsidised private cars by building and maintaining highways while neglecting means of public transport. This factory was mass-producing Plymouth cars in the 1940s.

116. **Urban sprawl.** The development of personal transport also made possible suburban living for the average family in developed nations. The proliferation of private cars and road networks gave rise to sprawling suburbs surrounding large cities everywhere, as extensive tracts of natural ecosystems and productive farmland were replaced by homes, streets and gardens. This is Hendon, a suburb in north London.

183

117. **Rising expectations.**
In poor countries, people
are often well informed
about the wealth of people
in rich countries, and they,
reasonably enough, hope
and expect to achieve a
similar affluence –
including the use of health-
threatening products such
as cigarettes. This ad was
photographed in a street in
Calcutta, India.

limited to risking black lung disease in coal mines or cancer from exposure to various toxic chemicals on the job, or being out of work.

Consequently, the people who controlled industrialised societies, even if they were humane and broadly concerned about the human condition, tended to have a distorted view of the distribution of costs and benefits in that society. And virtually everyone except trained ecologists overlooked the less easily perceived costs – the dissipation of humanity's priceless capital of natural resources, and in particular, the related accelerating deterioration of ecosystems and the life support services that they provide.

118. **Las Vegas lights.** A monument to human folly is Las Vegas, a city built in the midst of a desert, where people go to play and risk their hard-earned money. As a symbol of the dissipation of humanity's capital, Las Vegas is almost without peer; the energy that powers the lights is lost to other, more productive uses, as is the water drawn from the nearby Colorado River to water the lawns of Las Vegas residents.

Small wonder the leaders of the world's dominant culture could view rapid growth of the human economic enterprise as almost an unalloyed blessing. They could easily believe that the history of humanity since the industrial revolution, and especially since World War II, was *the* history of importance. They could see no reason why the trend would not be forever upward to more and more affluence for more and more people.

So the warnings of Plato, Chief Sealth, Thomas Malthus, George Perkins Marsh, Charles Elton, Aldo Leopold and others not only fell on deaf ears, they were interpreted as wrong by the short-term minds of economists and politicians. Humanity continued to rape and plunder the planet, and it had a larger population size and more prosperous people than ever before. This proved, especially to the frontier mentality of Americans, that the planet was essentially infinite – that the raping and plundering could go on forever.

Nowhere has the failure to recognise folly been more obvious than in the persistence of one of the most pernicious of all delusions – economic growth-mania. This is the widespread notion that there has always been economic growth and there always will be economic growth. As a British economist, Wilfred Beckerman, put it, since economic growth has gone on 'since the days of Pericles,' there is 'no reason to assume it cannot continue for another 2,500 years'.

119. **New York Stock Exchange.** Perhaps even more fervently than professional economists, businessmen believe in the impossible – the perpetual expansion of national economies.

A social scientist, Jack Parsons, put this remarkable idea to a mathematical test. It turns out that if economic growth had occurred at the very modest (from the viewpoint of economists) rate of one per cent a year from the time of Pericles (490–424 BC), the yearly income of the average family at that time would have to have been far less than a millionth of a penny!

Projecting growth rates forward at the same rate gives equally preposterous results. Two and a half millennia from now, the *per capita* gross national product of Britain would be about £400,000,000,000,000 – that is, the average Englishman would have about 200 times the wealth of the entire United States population today.

Should this little exercise in arithmetic be brought to the attention of an economist, one of two statements will be forthcoming. One is that the marvels of science have removed all constraints on economic growth, so even if it has not been continuous in the past, it can be continuous in the future. (Who knows, maybe the British will have found marvellous new ways to accumulate and dispose of wealth.) The other response is somewhat more intelligent, although (as you can see from what has gone before) still totally incorrect. That response is that indeed there are limits to economic growth, but they are at least hundreds of years in the future, so they need not concern the present generation or those immediately following.

Many economists simply *believe* that science can always pull a technological rabbit out of a hat to solve any problem. They don't understand that there are very important laws of nature that limit what kinds of rabbits can be pulled out of which hats, and so persist in assuming, among other things, that perpetual motion machines will soon be with us. And they pay no attention whatever to the additional fact that many of the rabbits that *can* be pulled out of hats produce such noxious droppings that it would be better if they had never left the hat in the first place.

The misunderstandings about how the world works that afflict cornucopian economists and the businessmen, politicians and magazine and newspaper editors who love to spread their soothing gospel are so profound that they can probably be cured only by intensive courses in basic mathematics and science. One example can serve for all. An American specialist in mail-order marketing, Julian Simon, has actually concluded that resources are infinite and therefore, the human population can presumably grow to infinity. This discovery was made with the help of an elementary mistake in mathematics and a convenient ignorance of the nature of resources.

The mathematical mistake consisted, in essence, of concluding that, since a line of any length can be divided into an infinite number of points, every line therefore must be infinitely long. Simon extended

187

this mistake to infer that any resource can be subdivided indefinitely and thus must be infinite in quantity. This of course would not be true even if resources could be subdivided indefinitely. But they cannot. If petroleum were subdivided far enough, its molecules would be broken up and the fragments would no longer be petroleum. If copper were subdivided at the atomic level, the fragments would no longer be copper.

120. **The Scopes trial.** In 1925, Tennessee high-school teacher John Scopes (centre) was convicted in the famous 'monkey trial' of teaching his students about the theory of evolution, which had been made illegal by a state law. Scopes is flanked by his attorneys, Dudley Malone (left) and Clarence Darrow, who headed the defence. Fundamentalist Christians in the United States have continued to oppose the teaching of the facts of evolution in public schools ever since, and have had considerable success in suppressing it. Most Americans accordingly lack a fundamental understanding of the biotic systems of which they are a part. In the Soviet Union, the anti-evolutionists triumped completely under Stalin – which is one reason why Soviet agriculture is in such disarray today.

In what must be one of the most unscientific statements ever published, Simon declared:

> Copper can be made from other metals. . . . Even the total weight of the earth is not a theoretical limit to the amount of copper that might be available to earthlings in the future. Only the total weight of the universe . . . would be such a theoretical limit.

Simon's confusion can be traced to the nuclear reactions in which one element can be changed into another. Copper can indeed be made from other metals – in fact, from a form of nickel. Unfortunately, though, because of the way the universe works, the process is doomed to be forever extremely expensive. For instance, one pair of physicists, enchanted by Simon's statement, calculated that, under exceedingly optimistic assumptions, it might be possible to produce copper from nickel at a cost of only a billion dollars a pound!

Simon's thesis does provide the basis for an amusing fantasy. Picture Simon with his magical copper-converter, an all-copper machine that is busily turning the entire universe into copper. The machine would have to be powered by some extra-universal energy source, since all of the energy in the universe is part of the mass that would have to be converted into copper to reach Simon's goal. He then jumps in front of the device and is himself converted into copper. At that point, his theoretical limit of the amount of copper that can be made available to humanity would have been reached – although, of course, all of humanity would have been converted into copper by then also, since we are all part of the mass of the universe.

Sadly, Simon's statements were published in the journal *Science*, the pre-eminent North American scientific publication. And a book offering similar examples of analysis of population-resource-environment questions was published by the prestigious Princeton University Press. Simon's views have also been taken seriously by the *Washington Post*, the *Wall Street Journal* and airline magazines.

Most of this can just be chalked up to the abysmal ignorance of everything related to science that pervades the American public – a result of the steady deterioration of science teaching in schools as people who would like to dedicate their lives to educating our young find themselves forced to choose making a living wage over that ambition.

The appearance of Simon's decidedly unscientific views in *Science*, however, illuminates another problem. Although the journal is putatively refereed, it is crystal clear that no referee who has ever had any training in science ever read Simon's paper from end to end. How, then, was it published? The answer would seem to be that many scientists are members of the establishment and are naturally delighted with anything that says the world is going in the right direction, and especially with anything that implies that science is what is keeping it on that road. Articles suggesting that science is not going to solve the human predicament, or worse yet, that it might be a contributor to it, are received with little enthusiasm by establishment journals.

Simple solutions to the human predicament have been embraced by some decision-makers with the same eagerness with which young children embrace the notion of the Tooth Fairy. Wealth will be placed beneath our pillows if we just keep doing what we've always been doing. No uncomfortable changes in our own behaviour will be necessary – economic growth (or shedding milk teeth) is all that is required. The beliefs that population growth can go on forever (or at least for a very long time), that all energy problems will be solved by nuclear fission or fusion power, that environmental difficulties are all relatively trivial issues of 'pollution', and that international security can be obtained by simply building more and more nuclear weapons, have been so appealing to the poorly informed citizens of the United States and many other Western countries that they have verged on being the majority view.

Nevertheless, despite the strength of the growth-manic myth, by the late 1960s, severe pollution problems, exploding populations in

121. **A Vietnam battlefield.**
The justifications for the
involvement of the United
States in Vietnam were to
protect its interests in
resources in the region
(which included petroleum,
tin and other minerals) and
to halt the spread of
communism. The resources
that were consumed or
wasted (not to mention the
human lives) in conducting
that unpopular war,
however, may well have
outweighed those that
might have been acquired.
These soldiers took part in
the battle for Hill 875 in
October 1967.

developing countries and a nasty famine in India convinced increasing numbers of people that humanity was not necessarily on an ever onward, ever upward course. This feeling was enhanced by the growing realisation that the rich nations of the world were quickly gaining the capacity to end civilisation with thermonuclear weapons. In the West, the environmental and anti-Vietnam War movements posed a major threat to those who thought that the 'free world' could grow forever economically, assuring its access to needed resources with military might and protecting itself from attack by the threat of massive nuclear retaliation.

By the 1970s, mounting evidence of a deteriorating population, resource and environment situation, highlighted by the energy crisis, the collapse of the American adventure in Vietnam and destabilisation of the arms race, made clear the need for a very careful reassessment of the assumptions under which all societies were run. And there was some sign of movement in that direction, especially during the Carter administration in the United States.

But the forces of reaction were too strong. A counter-attack was launched in the West designed to keep humanity on the course toward destruction. There was a major victory for those forces when Margaret Thatcher became Prime Minister of the United Kingdom. Shortly thereafter, in 1980, the United States may have made its terminal mistake: it elected Ronald Reagan as president.

122. **Prime Minister Margaret Thatcher and President Ronald Reagan.** These two leaders share a philosophy of government, economics and foreign relations that might have been appropriate in an earlier time but is dangerously obsolete in the age of nuclear weapons, resource depletion and environmental decay. Rather than bringing back the exuberance of the nineteenth century, they may be harbingers of global poverty in the twenty-first – if not of postnuclear – devastation. In December 1984, they shared a jolly ride in a golf cart at Camp David, Maryland.

The Reagan administration was swept into power by a public that was tired of hearing about problems and desperately wanted to believe in the Tooth Fairy. The result was to empower a government suffused with nineteenth-century attitudes and utterly ignorant of all of the major issues facing humanity in the last part of the twentieth century. At the most critical juncture yet in human history, the most powerful nation in the world had taken a perilous turn.

The Reaganites were enormously effective at developing and instituting retrograde policies in areas as diverse as population, the environment, civil rights and the arms race. Consider the Reagan administration's behaviour relative to the critical issue of population control. Backed by fundamentalist religious sects, it attacked family planning programmes and abortion rights on a variety of fronts. Few if any ethical issues have been as divisive and potentially disruptive as abortion. The issue deserves careful scrutiny, as it illustrates the folly of politicians depending on simplistic beliefs rather than informed analysis.

Abortion is one of the most widely practised methods of birth control – an estimated 50 million abortions occur worldwide each year (compared to about 130 million live births). The only effective

191

techniques more widely used today are steroid pills and sterilisation. Like contraception, the technology of abortion has advanced to the point that it is safer than childbearing (a situation that did not prevail a century or so ago when religious opposition to abortion originally developed). Yet here too there is room for further improvement, especially through research on 'morning after' pills (basically chemicals that cause early abortions).

Today, abortion in early pregnancy is legal under a wide range of circumstances in the majority of nations, which contain two-thirds of the world's population; elsewhere it is relatively restricted or completely illegal. But in some nations such as Britain and the United States, where abortion is freely available, the procedure is deemed immoral – equated with murder – by a small but vociferous minority of the population. The Reagan administration chose to ally itself openly with this minority view.

The fundamental question in the debate is often framed as: 'When does human life begin?' The anti-abortionists' answer is: 'At the moment of conception.' Biologically, however, the question makes no sense; human beings, like most eukaryotic organisms, exist as a continuous alternation of two stages of being. In one stage (simplifying slightly), chromosomes within unspecialised cells exist as nearly identical pairs. The chromosomes are microscopic structures that carry most of the DNA, the giant molecules into which are coded the information necessary for carrying out life processes, including reproduction. Each unspecialised cell in your body carries 23 *pairs* of chromosomes – a total of 46. In the alternate stage, that of reproductive cells (sperm or egg), each cell contains just one member of each chromosome pair. Each human egg and sperm cell thus contains 23 chromosomes.

Biologists speak of the paired chromosome stage as the 'diplophase' of the life cycle, and the unpaired stage as the 'haplophase'. Fertilisation of an egg marks the transition from haplophase to diplophase, while a specialised kind of cell division that produces sperm or eggs marks the return of diplophase to haplophase. In both phases, there is considerable natural 'wastage' of individuals – the overwhelming majority of sperm and most eggs die without giving rise to a diplophase individual. Numerous diplophase individuals die before producing eggs or sperm (a great many embryos are naturally aborted before a woman even realises she is pregnant, and many are miscarried in later stages). Until recently, infants born prematurely rarely survived. Indeed, before this century, a large proportion of full-term babies and children died before maturity. In many poor countries, they still do.

Since the human beings who discuss moral issues or take part in anti-abortion protests are inevitably in the diplophase, it is not surprising that the biologically uninitiated tend to confuse the beginning of the diplophase with the beginning of life. But that is biological nonsense; an egg or sperm is every bit as 'human' as you or we are. Among some other organisms such as mosses, the haplophase is the prominent, obvious phase of the life cycle (the leafy moss plant is the chromosomal equivalent of an egg or sperm; its inconspicuous, dependent reproductive part is the equivalent of an adult person). The bottom line is that, when life begins (or, more correctly, when a person to whom we impute rights starts to exist) is a legal and moral question that cannot possibly be answered biologically.

123. **An anti-abortion demonstration.** These demonstrators were holding a protest rally outside a Planned Parenthood clinic in Philadelphia in 1985. Many clinics in the United States have been picketed by demontrators who harass clients visiting the clinics. More ominous, dozens of clinics have been bombed by anti-abortion extremists; so far no one has been seriously injured or killed. Unhappily, most anti-abortionists are uninterested in supporting the one enterprise that could make abortion largely obsolete – making safe, effective contraception available to all sexually active individuals.

That is not to say that it is not an important question for society – it clearly is. For a long time, an individual was defined legally as coming into being at birth, at a time when it could survive outside its mother's body. But advances in technology have made such a definition increasingly obsolete. With the steady improvement in methods of maintaining premature infants, many that certainly would have died twenty years ago can now be given a reasonable chance of life. Indeed, no theoretical barrier seems to foreclose the eventual option of developing eggs fertilised in a test tube into children entirely outside a human mother. It won't happen this year, but few biologists would declare it impossible.

Perhaps, then, the anti-abortionists have hit on an appropriate point at which to define the start of personhood – the 'moment' of fertilisation (a moment difficult to define, since fertilisation itself is a rather complex sequence of events). Yet this definition presents a new set of problems. Some contraceptive techniques, such as IUDs, may work by preventing implantation of the fertilised egg and thus would become devices for 'murder' if that egg were granted full human rights.

It is worth pointing out, in addition, that after birth, children are not granted full rights by any society, and decisions regarding their welfare are normally left to their parents. The rights and responsibilities relegated to parents are usually assumed by society only in cases of gross negligence or abuse. And the trend in this century has been to affirm the right of parents to decide the number of their children. Thus the greatest problem with making abortion illegal again is that it would condemn millions of women to bear children they do not want, and would affix the label of murderess on millions of others who would resort to abortions regardless of whether they were legal or illegal.

Professor Carl Djerassi of Stanford University has commented on the likely consequences of making abortion illegal in the United States and elsewhere, citing the cases of Ireland, Spain and Romania. Ireland and Spain are developed countries with literate populations in which abortion is illegal. That does not mean, however, that Irish and Spanish women never have abortions. In 1981, a typical year, some 3,600 Irish women and 20,500 Spanish women travelled to England for legal abortions – a significant proportion of the women of reproductive age in their respective nations. Of course, these statistics underestimate the frequency of abortions among Irish and Spanish women, since they do not include illegal abortions obtained within those nations or legal ones obtained elsewhere.

More insight comes from the experience of Romania, where until 1965 abortion was legal and easily obtainable. Then the government, concerned about a falling birth rate, suddenly introduced restrictive abortion laws. There was an immediate, dramatic rise in the birth rate the next year, followed by a nearly equal drop the year after. There was no change in the availability of contraceptives it had just taken a couple of years to establish an illegal abortion network. The main impact of Romania's outlawing of abortion was a tenfold increase in the maternal death rate associated with abortions.

There is every reason to believe that the reinstatement of restrictive abortion laws in the United States (or other countries) would have the same result – a small reduction in the number of abortions accompanied by a large rise in the number of women dying from bungled illegal ones. Today's campaigners for strict abortion laws have forgotten (or are too young to know) that there were an estimated one million illegal abortions in the United States annually before the laws were liberalised beginning in the late 1960s. That is almost as many abortions as occur now, and there were far fewer women of reproductive age in the population then. But, of course, the maternal death rate (not to mention rates of injury and infection) was horrendous, whereas now abortion is one of the safest medical procedures.

Is there an ethical way out of the abortion dilemma? One that is recommended by 'right-to-life' advocates is that the unwanted babies be carried to term and then put up for adoption. Suppose virtually all the women with unwanted pregnancies made that choice. Statistics show unambiguously that the demand for babies for adoption is insufficient to absorb more than a small proportion of the unwanted babies that would be born. In addition, sociological studies in European countries have indicated that children born to women who were denied abortions often have more probems in school and in later life and are more likely to become burdens on society than are children from similar families who were wanted.

The ethical way out of the abortion dilemma from the standpoint of freedom of choice is simply to leave the moral decisions to the people most closely involved, primarily the pregnant woman, her husband or partner, and her physician. But this position does not take into account the feelings of a segment of society who are deeply offended by the thought that abortions are taking place. An ethical decision should, as far as possible, take into consideration the views of minorities. In this case, the solution that should be satisfactory to both sides lies in education about and dissemination of birth control devices so that every sexually active individual is willing and able to control his or her fertility.

Unfortunately, some of the most adamant objectors to legalised abortion are also opposed to birth control, and especially to providing information and materials to unmarried teenagers, on the questionable grounds that it leads to promiscuous behaviour. Individuals have a right to hold such a self-defeating position, but it is not shared by the majority. None the less, elements in the Reagan administration and right-wing conservatives in Congress have attempted to enshrine it in national policy. Not only have they tried to repeal legal abortion and succeeded in restricting the activities of private family planning organisations within the United States, they have endeavoured to impose their moral views on people in other countries.

Perhaps the low point of the Reaganite assault on the notion of population control came in 1984 at the United Nations Conference on World Population in Mexico City. The Reagan administration had

195

picked as one of its population gurus the very same Julian Simon who looks to alchemy to expand the supply of copper. Another was journalist Ben Wattenburg, now with the American Enterprise Institute, who favours an increase in the American birth rate out of fear that the next generation of young men will be too small to defend the nation adequately, or to pay taxes to support the burgeoning nuclear armamentarium. Apparently, the possibility that hordes of cannon fodder might be useless in an era of automated battlefields (should the developed nations indulge in the ultimate folly) has never crossed his mind.

In Mexico City, the result was low comedy – the United States government took an ancient Maoist position at the conference. Chairman Mao used to emphasise that 'of all things, people are the most precious' when he opposed any steps towards population control; Simon writes about people being 'the ultimate resource'. The old Chinese communist position was that if a nation had the proper economic system, its population problems would take care of themselves. The Reaganite position was essentially the same – differing only in its choice of economic system that would allow the laws of nature to be broken. The end result was that the United States government looked silly in a world forum, but happily, the delegates from other nations did not take the Reagan-Simon view seriously.

Unfortunately, the American right wing did not just push the government to make a fool of itself in Mexico City. It has also attempted to cripple American assistance to people in poor nations who wish to limit their families. Congress so far has stood more or less fast in maintaining foreign assistance for population programmes, but the administration nevertheless withdrew funding for the International Planned Parenthood Federation in 1985, and the next year cut off US support for the United Nations Fund for Population Activities (UNFPA) on the grounds that these organisations provide abortions. Both organisations at most support counselling programmes, and funds for those have been kept in strictly separate accounts following earlier US legislation forbidding the use of American foreign aid monies for abortions.

Serious as the Reagan administration's attempts were to promote the population explosion, its efforts to dismantle hard-won gains in environmental protection have been even more successful. The White House Council on Environmental Quality was all but destroyed, removing in one stroke the nation's environmental conscience and leaving behind an ineffective shell. The Environmental Protection Agency (EPA), once widely cited as the most effective regulatory agency in the federal government, was put in the hands of good friends of polluters and its funds drastically cut.

When the public uproar, especially about toxic wastes, led to the removal of some of the worst administrators in the EPA, an attempt to restore some level of respectability was made by appointing William Ruckelshaus as director. This move more or less succeeded in its limited goal, but restoration of the agency to its former regulatory effectiveness will require many years of adequate funding to rebuild an administrative and scientific staff with the necessary experience and expertise.

In the long term, the most damaging environmental actions taken by the Reagan administration may prove to be the changes made in agencies responsible for management of the land and other resources of the United States. The administration carried out a campaign to maximise the reckless exploitation of resources in this generation, with little consideration either for the ecological costs or the consequences of such a policy for future generations.

The campaign was launched with the appointment of the flamboyant James Watt as Secretary of the Interior, but many quieter appointees in other positions have helped move the nation towards the drain. The Forest Service emphasised exploitative policies, and the BLM (Bureau of Land Management – which might better be named the Bureau of Logging and Mining) redoubled its efforts to ensure that no resources would be left for the future. Rather than meeting their legal obligations to reverse destructive environmental trends in the United States, the Reaganites did their best to accelerate them.

Unfortunately, many of these trends were invisible to the voting public. One example is the steady degeneration of public lands in the western United States (where about half the land is owned by the government), largely at the hands of a segment of the cattle industry that makes an insignificant contribution to America's beef supply.

Over-grazing, mainly on federal lands, is causing desertification at an alarming rate in the Great Basin, the region between the Rockies and the Sierra Nevada. Moist areas along streams are often particularly hard hit, for there the graze is most abundant. Cattle mow down the plants, trample the stream banks and pollute the water with their wastes. The destruction of shade heats up the streams, which, combined with erosion and siltation from the damaged banks, wipes out valuable fishes such as trout.

It is a game in which nearly everyone loses. Hunters and fishers find less game and fish, campers are often greeted by campgrounds that are wall-to-wall cow pies and must drink and bathe in polluted water, while municipalities and industries lose clean water supplies. The beneficiaries of this destruction are fewer than 35,000 cattle ranchers, a politically powerful group, and the congressmen and bureaucrats who do their bidding. The whole cattle operation on the public lands of the west supplies less than 5 per cent of American beef production, though

197

about a quarter of the grazing land in the United States is used to achieve that puny goal.

The final irony is that Americans have learned that eating too much red meat is not good for them and have reduced their consumption of beef. The western ranchers and their government cronies are destroying a gigantic portion of the United States to add a few percentage points to the production of a commodity that the industry is now desperately trying to promote in an era of declining demand.

It would be an enormous bargain to pay each of the 35,000 ranchers a million dollars simply to cease and desist – 35 billion dollars (one-sixth of one year's defence budget) would be a tiny price to pay to prevent the desertification of a large part of the nation. Another solution, of course, would be to regulate much more carefully the areas grazed and the stocking rates so that a sustainable cattle industry and the much admired way of life associated with it could be preserved

124. **Short-term agricultural production.** An example of the foolish burning of humanity's capital is this series of irrigated wheat fields in western Nebraska, an area with too little dependable rainfall to support high-yield agriculture. The irrigation water is pumped from the giant Ogallala aquifer which underlies much of the high plains region of the United States. In many areas of the Great Plains, water is being pumped out of the aquifer at a much higher rate than it can be recharged. High-yield agricultural production is possible under these circumstances for only a few decades.

permanently. Properly managed, such a programme might even reverse the desertification of the intermountain west and begin to restore the original grasslands.

The sprint of folly has by no means been limited to the United States, of course. Prime Minister Margaret Thatcher of the United Kingdom evidently shares President Reagan's lack of enthusiasm for environmental protection. While the British government has chalked up some success stories such as the return of salmon to the Thames, in many other ways it has lagged behind the United States in implementing pollution control measures and safeguarding natural resources for the future.

For instance, the herbicide 2,4,5-T, which contains trace amounts of the highly carcinogenic (cancer-causing) compound dioxin, is still widely available and used in Britain. It was dioxin that caused the abandonment of the little town of Times Beach, Missouri, in the early 1980s after its roads were resurfaced with contaminated material, and that made the area around Seveso, Italy, uninhabitable after an explosion in a chemical works in 1976. An explosion in a factory in Derbyshire in 1968 had roughly the same results, but no action to control the dioxin-containing herbicide has ensued. Similarly, the British government, along with the rest of Europe, has dragged its heels in removing lead from petrol, despite abundant evidence of its adverse health effects.

Along with other industrialised nations, Britain has begun to face the problem of long-ignored toxic wastes that have escaped their disposal

125. **Salmon caught in the Thames.** In 1983 the first salmon caught for over 150 years with a rod and reel in the Thames near London was displayed as proof of the successful clean-up of the river.

areas to contaminate food and water sources. As of the mid-1980s, no national policy had been developed for toxic waste management, including the management of radioactive wastes. Leaks and discharges of radioactive materials from the Windscale nuclear power plant have been an ongoing scandal since the 1950s – a stink that was not abated by changing the plant's name to Sellafield.

Deceiving and misinforming the public about the dangers of leaks and accidents at nuclear power plants, however, is a tradition that many governments and the nuclear industry have honoured, from Windscale in the United Kingdom, to Three Mile Island and several less well-known events in the United States, to the most recent disaster at Chernobyl in the Soviet Union. Ironically, however, the Soviets encountered a storm of criticism from Western governments when they were slow to release information on the accident.

Opposition to nuclear power in general, and specifically to operation of plants of doubtful safety, rose dramatically after Chernobyl in both North America and Europe. In the United States, the ineptitude of the nuclear power industry – especially the premature deployment of an inadequate technology – together with the string of publicised accidents, had already been threatening to remove fission power from the array of potential future energy sources. It is conceivable that reactors could be developed that would be secure against both catastrophic accident and contributing to the spread of nuclear weapons. And the problem of disposing of nuclear wastes might one day be solved. But officials have lied to the public so often and so blatantly that it might well refuse to support any further expansion of fission power, however much improved.

The history of control of acid rain is equally dismal on both the North American and Europēan continents (including large offshore island kingdoms). National leaders have wasted years of valuable time pointing fingers of blame at each other and insisting that more research is necessary before polluters can be pushed to remedy the problem. Meanwhile, many forest and fresh water resources are being swiftly and irreversibly destroyed, but not so fast that most of the politicians won't be out of office before the devastation is complete.

The future of *Homo sapiens* is seriously threatened by the consequences of our species having overshot the long-term carrying capacity of the planet. But the future is also shadowed by population-related increases in international tensions. Squabbles over declining resources, trans-national environmental problems, the migration of economic and political refugees and different rates of population growth of ethnic groups within nations all enhance the prospects of military conflict. And that, in turn, increases the chances of a nuclear war.

A large-scale nuclear war (the only sort experts believe likely) has a high probability of being followed by a nuclear winter. In that event, enormous amounts of soot and smoke from bomb-ignited fires would dim the sun and lower temperatures over large areas of Earth for many days to many weeks. At the least, these effects would severely damage agriculture in most of the Northern Hemisphere, causing the starvation of hundreds of millions of people who had survived the initial blast, heat, fires and prompt radiation. At worst, if nuclear arsenals continue to grow, these effects and other grim ecological consequences of the war could threaten the destruction of many of Earth's ecosystems and possibly the survival of humanity itself.

The ultimate folly of twentieth-century humanity must therefore be the increasingly destabilised arms race between the superpowers. The state of armament was such by the mid-1980s that, in time of crisis, military advisers on both sides would be forced to press their governments to 'shoot first'. This situation primarily traces back to the decision by the United States to 'MIRV' its intercontinental ballistic missiles. MIRVing means that each missile has more than one warhead, each of which can be aimed at a different target.

The United States went ahead with MIRVing for the customary reason – it would keep America technologically ahead of the Soviets. It was one of many examples of the 'fallacy of the last move' which is a driving force in arms races: the idea that, if your side gets some new weapon, the other side will be unable to catch up or counter it. As expected by all sensible analysts, the Soviets quickly caught up in MIRVing.

126. **John Wayne.** One of two great symbols of American machismo and proponents of a belligerent national posture, especially toward communist states, was actor John Wayne; the other is Ronald Reagan. Neither ever saw action during wartime. This photograph is from the 1946 film *Red River*.

The reason that MIRVing is destabilising is simple. Suppose each side had 1,000 missiles, each MIRVed with ten re-entry vehicles (nuclear warheads and the shell that protects them as they re-enter the atmosphere after their brief trip through space). In their silos, they represent 1,000 targets, each of which, for the sake of illustration, might be destroyed by two warheads from the other side. To send the requisite 2,000 warheads, one nation need only launch 200 of its 1,000 missiles – leaving it with 800 missiles (8,000 warheads) after all those of the other side were destroyed. MIRVing thus gives a great advantage to the side that shoots first, at least in the minds of conventional tacticians. Of course, neither side could really destroy all the other side's nuclear weapons, and those that remained – on missiles in silos, on cruise missiles, on bombers, or in nuclear submarines – would be more than adequate to send the 'victor' back to the Stone Age.

The Reagan administration's insane promotion of nuclear war has accelerated the pace of the arms race and contributed to further destabilisation. Reagan, and especially his Secretary of Defence, Caspar Weinberger, have constantly reiterated the lie that the United States is

201

far behind the Soviet Union in armaments. In truth, the United States, because of its greater technological competence, has maintained the lead in most meaningful measures of military capability. This is clearly shown by the repeated testimony before Congress by the chiefs of staff of each of the American armed services that they would not trade their branch of the service for the Soviet counterpart.

Several other steps taken recently by the United States are also destabilising. One was the forward placement of highly accurate Pershing II missiles in Europe. To the Soviets, these can look like weapons designed to decapitate their command and control systems by striking hardened bunkers around Moscow with only a few minutes warning. Although US submarine-launched ballistic missiles (SLBMs) could hit Moscow with similar warning times (as Soviet SLBMs could hit Washington), those missiles do not now have the accuracy to dig out well-shielded military nerve centres. Since Soviet military planners, like ours, must continually make 'worst case' assessments, they cannot help but feel that the Pershing II missiles in Europe could make a big contribution to a US first strike on their homeland. One can well imagine the response of the US military if an equivalent force of Soviet missiles were emplaced in Cuba.

Another step was the decision to place MX missiles in silos that once held Minuteman missiles (the mainstay of US ICBM forces in the 1970s and early 1980s). The MX represents a big 'advance' over the Minuteman III (the most capable of the Minuteman series); it carries ten warheads per missile as opposed to three for Minuteman III. Both its range and accuracy are also greater, a highly significant matter when it comes to destroying hardened targets such as missile silos.

The original US plan was to deploy the MX in a sort of 'shell game' in the deserts of the intermountain west so they could survive a Soviet first strike and be available to retaliate. Now the missiles are to be deployed in silos whose positions have been in the enemy's targeting computers for more than a decade. Again, the Russians must suspect that such vulnerable basing indicates an American first-strike strategy, since most of the MXs would not survive a well-orchestrated attack launched by the USSR.

In 1983 the Strategic Defence Initiative (SDI), the 'star wars' defence, was proposed. No serious analyst believes that, even if such a system could be designed and it worked well (most scientists believe it would not work at all), it could possibly protect American cities from a determined Soviet attack. There are simply too many ways to counter the SDI system, or to avoid it entirely by sending thermonuclear bombs in airplanes, cruise missiles, submarines (underwater detonations off the US West Coast could kill huge numbers of people with fallout) or suitcases, against all of which SDI is powerless.

127. **MX missile launch.**
The giant MX missile is the newest in the United States' nuclear arsenal. Each one is capable of carrying ten independently targeted warheads which can strike with impressive accuracy. Because of their capacity to destabilise further the arms competition, deployment of these missiles is likely to reduce substantially the military security of the NATO alliance. This photograph shows the missile's 'maiden' test flight in June 1983 from Vandenberg Air Force Base in California.

But if SDI did work, it might be able to prevent a large proportion of the nuclear warheads in a Soviet retaliatory strike from reaching their targets in North America. There is every reason to believe that, even in a first strike, many Soviet missiles would misfire or go off course; the ragged return volley that the Soviets might be able to launch after a massive attack from NATO should prove much simpler to defend against. Thus the Soviets are likely to view SDI as another element in an American first-strike strategy.

Much of the American public has been persuaded that SDI is not only possible, but that it offers real protection of the population against a Soviet attack. The Reagan-Weinberger charade must be a near record for pulling the wool over the eyes of a public that has ample information to the contrary available to it — if it only chose to use it. In its defence, however, the public ought to be able to trust its leaders to tell the truth.

People in the West should remember that the Soviet Union is both powerful and paranoid. We do not believe that the United States is planning a first strike; but the American strategic posture (and the rhetoric of the conservative elements that came into prominence in the Reagan administration) has increasingly led Soviet planners to fear that

203

is exactly the intention. As a result, the Soviets may have moved towards a 'launch on warning' or 'launch under attack' posture. This means that, in time of crisis, the decision on whether or not the Soviet Union is under attack and should retaliate would be left largely to computers, since there would not be enough warning time to involve people.

When one considers that sophisticated US computers charged with evaluating radar and satellite intelligence on Soviet military operations have frequently and repeatedly issued false warnings of a Soviet strike, this prospect is most unsettling – especially so because computer technology is one of the areas where the Soviets are furthest behind the US.

So folly piles on folly, and Earth is threatened with both a steady deterioration of its life support systems and the possibility of the massive destruction of a thermonuclear holocaust. Yet, in spite of the seemingly inexorable sprint of folly, some trends do give us cause to hope that this time humanity may alter course before it is too late.

128. **Last launch of the Challenger Space Shuttle.** The explosion of the Challenger space vehicle in January 1986 demonstrated once again that fools of sufficient magnitude can be found to overcome any 'fool-proof' system. Complex technologies always carry risks of failure. To gamble the entire future of civilisation on a technology of unprecedented complexity and unproven effectiveness, as the Reagan administration wants to do with its Strategic Defence Initiative, is the height of folly.

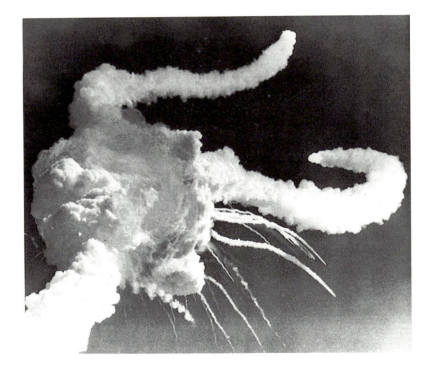

CHAPTER NINE

The News is not all Bad

Earth is being battered by *Homo sapiens*, but it is still capable of recovering if its 'opponent' gives it half a break. There are some hopeful signs of that happening. More and more people have begun to recognise that our species is collectively headed for a cliff. And not all aspects of the human condition point toward an inevitable catastrophe. Above all, there are numerous signs that under the right circumstances, societies can radically and quickly change their attitudes and behaviour.

One of the best examples of such a change was the dramatic decline in birth rates in the United States in the late 1960s and early 1970s. In the 1960s, most social scientists believed that at least several decades would be needed for the United States population to reach 'replacement reproduction' – the level of fertility at which couples had on average just enough babies to replace themselves.

But that point was actually reached in the United States around 1972, and fertility continued falling to below replacement level, where it has remained. A mix of factors was apparently responsible for this rapid change, including economic conditions not conducive to formation of families, the rising women's liberation movement and general concern over environmental deterioration and its connection with overpopulation. Americans were worried about the kind of country their children would inherit if the number of people in the United States continued to grow.

The decline in the birth rate was not a short-term phenomenon; in 1986, American fertility was still below replacement level. But the population was still growing, thanks to the built-in momentum of population growth and a high rate of net immigration. The low fertility persisted even though the basic policies of the United States government remained more or less pronatalist, especially in the pro-population growth years of the Reaganites. Clearly, important social and demographic shifts can occur regardless of governmental policies.

The existence of an array of safe modern contraceptive techniques has played a major role in permitting people, especially in rich countries, to regulate their fertility safely and conveniently. While unwanted pregnancies can be avoided, it is not entirely clear by how much the availability of dependable contraceptives has helped to reduce fertility, since to some extent they have substituted for abortion, withdrawal, abstinence and probably even infanticide.

For example, when the sale of contraceptives was illegal in Italy, an extremely high rate of illegal abortion kept the Italian birth rate in line with countries of similar economic status where contraceptives were available. Thus the impact of contraception on birth rates (as opposed to ease of regulating family size) may be overrated. But there is little doubt that birth rates are lower in many parts of the world today because contraceptives are available.

The technology of birth control presents few moral problems for most people in developed countries. Generally, the dangers of contraception are negligible for men and significantly less than those of childbearing for women. None the less, there is considerable room for improvement in both safety and convenience of contraceptive techniques. Advances are being made, but the research effort is small compared to what it ought to be – small, for example, compared to research on cancer, a problem of considerably less moment to the human future than that of population control.

Opposition on moral grounds to 'artificial contraception' by the hierarchy of the Catholic Church persists, but the ban on its use is widely ignored by Catholics. The 'rhythm method', sanctioned by the Church, has a much higher failure rate than other forms of birth control (more than five times that of the pill), due both to the unpredictability of women's ovulation cycles and the high degree of motivation and restraint required of its practitioners.

In the light of the potential consequences of continued population growth, religious opposition to artificial birth control must be viewed as a temporary aberration – a position that will not prevail much longer in the face of the threat of all that humane people hold dear. Instead, we hope that the practice of artificial contraception (including sterilisation) will be viewed as what it is biologically – one of the very few uniquely *human* practices, something that clearly differentiates people from other animals. No other species can knowingly limit its own reproduction. There is no ethical or religious middle ground on artificial contraception; its practice is the most moral of acts, since it is often taken out of consideration for the well-being of the next generation.

The availability of contraceptive methods is insufficient in itself to produce a reduction in population growth rates, as testified by the stubbornly high birth rates in many developing nations today. However, progressive government policies, combined with provision of the means of birth control, can produce impressive results.

In the 1970s, after two decades of vacillation on its official population policy, the leadership of the People's Republic of China began openly recognising the threat that overpopulation posed to their nation. The programme for family planning (called 'birth planning' in China) was already being widely implemented at the commune (village or urban neighbourhood) level through the health services, which provided information, birth control pills, IUDs and other contraceptives, voluntary sterilisation after completion of family and abortion, if needed. Communes were encouraged to regulate marriages and the numbers of children born in each one. Conformity to the goals of the birth planning campaign was largely generated through peer pressure in the intricate organisational structure of the communes and work brigades.

129. **A Chinese family planning poster.** No one who has visited China can doubt that the Chinese government is serious about its 'birth planning' programme, which has achieved remarkable success. This poster was photographed in the provincial capital of Sichuan in 1985. Note the emphasis on one child – a girl.

At the time of the first United Nations World Population Conference, held in 1974 in Bucharest, the Chinese 'birth planning' programme already was in practice by far the strongest national population programme in the world. Yet the Chinese delegation at the conference was still proclaiming the Marxist line that population problems were the result of a faulty economic system. By the late 1970s, however, the Chinese government's propaganda had caught up with its domestic policies. Conceding that rapid population growth had been hindering development, the official line was that the government's goal now was to end that growth and it advocated a two-child family as the norm.

207

Then China's leaders received a rude shock. Preliminary estimates prior to their 1982 census revealed that the population had reached one billion people before 1980, not a mere 900 million as had been assumed. The discovery meant that much of China's vaunted economic progress had been illusory. Production of food and goods had been measured in terms of production *per capita*; the Chinese suddenly realised that their divisor was wrong. There were 10 per cent more 'caputs' than they had figured on.

The government knew that something had to be done. After conducting a careful inventory of China's natural resources, the country's leaders concluded that the carrying capacity of the nation's territory was no more than 650 to 700 million people, if those people were to have a relatively good life. Therefore, for the first time in human history, a government launched a campaign not only to stop the growth of a country's population as quickly as possible but then to *reduce* its size – in this case, to lower the number of Chinese to fewer than 700 million.

Because of the inescapable momentum of population growth, the Chinese population was bound to expand well beyond 1 billion before its growth could be humanely ended. The Chinese goal was to contain it at a maximum of 1.2 billion, with the peak reached in the early twenty-first century. Many demographers doubt that the peak can be held below 1.5 or 1.6 billion, but are nevertheless impressed with the progress that has been made.

In the late 1960s, when China had succeeded in drastically reducing its death rate but the birth rate was still high, natural increase was about 2.8 per cent per year. By 1980, the Chinese claimed that their annual population growth rate was only 1.2 per cent (although Western demographers more conservatively estimated it to be 1.4 per cent). In a dozen years or so, the birth rate in Earth's largest population had been cut by half, an unprecedented accomplishment. By 1985, the average family size had fallen to replacement level – about 2.1 children per woman – and the annual growth rate was 1 per cent.

The stringent new population goals were to be met by strengthening the already existing birth planning programme, specifically by promoting one-child families for roughly half of the eligible couples (resulting, it was hoped, in an average family size of 1.5 children). Various incentives have been offered to couples to join the programme, including better housing and jobs, special educational opportunities for the children, social security guarantees for the parents and so forth.

Not surprisingly, in the course of such a vast social experiment, abuses occurred. In some cases, women have been more or less coerced into having abortions, and an increase in female infanticide has been reported. The latter is associated with the high value that Chinese culture traditionally places on sons. Even many progressive Chinese consider the birth of a female first child a disaster in a society that so strongly discourages second births.

To its credit, the Chinese government has been open about the abuses and attempted to deal with the problem by allowing second births for couples with daughters and under various other special circumstances. It has also attempted to counteract the preference for sons by emphasising the value of girls. And over-zealous local officials involved in coercion of abortions and sterilisations have been punished. Ironically, the revulsion that has been expressed by some right-wing American ideologues toward the Chinese one-child programme is largely based on information available because of the candour of the Chinese about their problems!

Currently, the Chinese government is considering a modification of the programme to allow second children at long intervals – eight to ten years – after the first. Thirty per cent of couples are expected to forgo the second child anyway, thus keeping the average family size at about 1.7 children. Stretching out the generation time with the long intervals between births will keep the growth rate low while smoothing out the sharp differences in generation sizes that would result from a one-child generation. China's goal of population shrinkage would take a little longer to achieve, but the transition would entail fewer distortions in the population's age structure. The current crop of babies, for instance, would not each be faced thirty or forty years from now with having to support two elderly parents with no siblings to share the burden.

While China is a trail-blazer in population control, other less developed countries have also struggled to curb their escalating human numbers. At the time of the first United Nations Conference on the Environment in 1972, many poor nations still had pronatalist and anti-environmental policies, reflecting the frequently expressed attitude that 'you've created your Los Angeles, Hamburgs, Birminghams and Tokyos, now it's our turn to destroy the environment for fun and profit'. The first United Nations Conference on World Population two years later was a highly politicised forum in which Third-World politicians accused the West of advocating population control for racist reasons. Some went so far as to label family planning aid as genocide.

But a mere decade later, these attitudes had largely disappeared as governments saw their nations' unemployment rates skyrocketing, their cities besieged and agriculture faltering, and began to realise that all their economic gains were being consumed by population growth. During the 1970s, the number of developing nations that still had no family planning programmes dropped to a tiny handful, and the claim that 'development is the best contraceptive' was no longer heard, as

209

leaders in poor countries saw that industrialisation did not automatically cause birth rates to fall, as they had been led to expect.

Indeed, demographers and sociologists have long debated over the causes of birth rate declines, both the early ones in the industrialised nations of the West and those occurring later in some (but not all) developing nations. The original assumptions that the processes of industrialisation and urbanisation were key factors to the demographic transition were given the lie by dozens of developing countries that had maintained high birth rates long after they had proceeded fairly far down the conventional development path. And some countries with high fertility rates also had relatively high per capita GNPs (relative wealth of the average person in monetary terms), which were also presumed to lead to low birth rates.

Conversely, a substantial decline in fertility occurred in some very poor nations whose societies were still predominantly agrarian in character – China being the most obvious case in point. Nor has the presence or absence of a family planning programme necessarily been a prime factor. Some programmes have met with success; in other countries, apparently equally strong programmes have been frustrated at every turn.

India, for instance, established the first national family planning programme in a less developed country in 1950. By the 1960s, the programme had strong government support and offered a variety of birth control methods, including voluntary sterilisation, IUDs, condoms

130. **Indian family planning elephant.** India has had a family planning programme since 1950, but, for a variety of reasons, has had much less success in reducing birth rates than has China. The elephant was employed to bring family planning information and materials to remote villages during the 1970s.

and birth control pills. Unlike some programmes, the Indian one made a serious effort to reach rural areas – no small task in that huge country, which has hundreds of thousands of villages, poor roads and several hundred languages. The population of India is the second largest in the world (behind China's), with about 785 million people in 1986. Yet, after decades of effort by the family planning programme, the Indian birth rate in 1986 was still stubbornly high – about 35 per 1,000 in the population – and the average family size was about 4.5 children. The annual rate of natural increase was 2.3 per cent.

In contrast, some other countries, no richer in per capita income and in some cases far behind in industrialisation, have succeeded in reducing their birth rates since the early 1960s, when the world population was growing most rapidly. In recent years, that success has led to more rapid gains in per capita GNP and accelerated progress in modernising the economies of those countries, but fertility had often begun dropping before the process had advanced very far. Some countries are still mainly agricultural. Besides China, these comparatively successful nations include Taiwan, South Korea, Panama, Colombia, Costa Rica, several Caribbean islands and Sri Lanka. Singapore and Hong Kong are special cases; they are almost entirely urban societies and today can hardly be termed 'less developed'.

Only recently has an answer to the demographic transition mystery emerged. The key to success in reducing birth rates is indeed related to development and modernisation, but evidently it has nothing to do with the number of factories a country may have or the proportion of the population living in cities. Rather, the essential factors appear to be low infant and child mortalities, a long life-expectancy, and education, especially of girls. When these have been achieved for the majority of the population, birth rates have often fallen, regardless of the levels of industrialisation or per capita GNP. A strong family planning programme and the availability of contraceptive methods can facilitate the trend, but without the key prerequisites, they may have little effect.

A little thought unmasks the reasons why these factors are important and others are not. Low infant mortality is related to the perceived values of children in traditional societies: as a source of cheap labour or family income, and as a form of social security. Obviously, if nearly all infants survive to adulthood, parents will not feel a need to replace children that have been lost, or to have several sons to ensure that at least one will survive to support them in their old age. Extended life expectancy is part of the same picture. Low infant mortalities and long life expectancies also imply reasonably high standards of nutrition, sanitation and health care.

211

131. **Child being immunised.** Saving the lives of children is an important prerequisite to reducing birth rates. The United Nations Children's Fund (UNICEF) reckons that three million children's lives could be saved each year if all infants were inoculated against such diseases as measles, polio, whooping cough, tetanus and diphtheria.

The importance of education for the modernisation of a society is fairly clear; without a cadre of workers with at least a minimal education, 'development' is hardly possible. On the other hand, governments, especially authoritarian ones, are wary of too much education – it breeds discontent. What has been less obvious is the importance of educating *women*. Indeed, in many traditional societies, especially those in which the status of women was low and pronatalist attitudes were strong, limited educational resources have been concentrated on boys, under the assumption that they, as future breadwinners, would need it most.

But it turns out that educated women, unlike men, apply their knowledge directly to the well-being of their families. Even a few years of schooling can result in measurably better nutrition, better health and better all-round care for children. This, of course, leads to lower infant and child mortalities and longer life-expectancies, which tend to increase population growth. But it also commonly leads to the adoption of family planning by individuals, which can more than compensate for the lower death rates.

Before these relationships had become clear, however, the majority of leaders in developing countries had seen the value in slowing rates of population expansion, even if most did not as yet accept the need

132. **Baby in basket.** This infant and her sister in Haranya, India, may face a brighter future if the world's governments can agree on the importance of limiting the human population size and building a sustainable world, and if they begin working together toward those goals.

for stopping growth altogether. One of the most positive global trends today is the increasing recognition by leaders of poor nations of the seriousness of the population explosion and the need to strengthen programmes to limit births. Thus the United States stood virtually alone when it took its antique Maoist-pronatalist stand at the United Nations World Population Conference in Mexico City in 1984.

A sign of the swift change in attitudes toward population control in developing countries could be seen in an appeal made in 1985 by the leaders of nations holding more than half of Earth's population. The heads of state of China, India, Egypt, Bangladesh, Kenya and thirty other nations signed a 'Statement on Population Stabilization', which was published in *The New York Times* on 20 October 1985. Among other things, they declared:

> Degradation of the world's environment, income inequality and the potential for conflict exist today because of over-consumption and over-population. If . . . unprecedented population growth continues, future generations of children will not have adequate food, housing, medical care, education, earth resources, and employment opportunities.

Pleading in essence for a change in the retrograde policy of the Reagan administration, they concluded:

> Recognizing that early population stabilization is in the interest of all nations, we earnestly hope that leaders around the world will share our views and join us in this great undertaking for the well-being and happiness of people everywhere.

We hope this statement will give pause to those in the United States government who have been listening to the claims of a small group of 'experts' that unrestricted population growth is beneficial.

Population size (and growth) is of course only one major component of the human predicament — only one of the three multiplicative factors that determine the length and depth of the human shadow across Earth. Also important is the behaviour of the people — the resources they use and how they treat their environment. Here, too, the record is decidedly mixed, but some good news can be seen.

In spite of the all-out assault on the environment launched by the Reagan administration, public concern for environmental quality in the United States has, if anything, continued to rise since President Reagan took office, as has been confirmed by numerous public polls. Indeed, public-interest environmental organisations prospered when the administration's scandalous policies were wrecking the government agencies responsible for environmental protection. Under pressure from Congress and environmental groups representing the public, the administration was forced to remove its most outrageous administrators and tone down its rhetoric. Its policies were not significantly changed; but backed by ample public support, Congress has been able to prevent much further damage.

Most Americans appear to view the Reagan policies as a temporary setback in an era of environmental progress, albeit a highly damaging one. Their commitment to environmental quality, combined with the emergence of Green parties in Europe and other Western nations, indicates that well-informed citizens in the rich countries will continue to press for needed action; the environmental movement was no flash in the pan.

There are also some encouraging signs that developing nations have begun to take their environmental problems seriously. Although air and water pollution are sometimes still viewed as welcome signs of 'progress', the tragic consequences of desertification, floods, droughts and growing shortages of fuelwood caused by deforestation, and declines in agricultural productivity resulting from soil erosion and poor management of irrigation systems are all clearly environmental problems that threaten the future well-being of the people involved.

133. **Tree-planting in Kenya.** In an effort to reverse the process of desertification, several nations of the Sahel have launched reafforestation schemes, often aided by international groups such as the US Peace Corps and Oxfam. Here a young couple water a seedling in Kalokol, in the Turkana region of Kenya.

In this context too, China is emerging as a leader. Its population policies grew out of a clear-eyed assessment of their territory's carrying capacity and a fair understanding of their environment's ability to tolerate human abuse — and it has been subjected to considerable abuse in 5,000 years. In our view, even 700 million people would probably exceed the long-term carrying capacity of China.

215

Only about 10 per cent of China's land is arable, and deterioration of that land (through erosion, salination of irrigated fields etc.) is a constant problem. Most of the nation's forests were destroyed long ago; the post-revolution Maoist regime virtually finished the job, adding drainage of ponds, lakes and wetlands to its record of destruction in a heroic effort to boost grain production to feed the huge, rapidly growing population. The grain production effort was successful (though some of the success is certainly attributable to success on the population front), but at enormous cost to the Chinese environment.

Indeed, the conversion of some lands to crops – especially remaining watershed forests, wetlands and fragile semi-arid areas – is now seen as a mistake that has cost more in productivity of cropland than was gained through increased cropland acreage. Even in Mao's time, the loss of the forests was recognised as too high a price, and a gigantic reforestation scheme was launched. Mao's more pragmatic followers have expanded the programme, but it is still inadequate in scale and too often based on exotic tree species, which cannot usually satisfactorily play the important ecological roles that native trees can.

Apart from putting unsuitable land to the plough, though, the Chinese approach to food production has generally been sensible – in many ways more sensible than the trends in the West – although, again, mistakes have been made. The first priority of the revolutionaries was to feed the population, much of which had periodically faced starvation. An early event in the Mao era, which only came to light decades later, was a massive crop failure, followed by a famine and the starvation of tens of millions of Chinese. No doubt stung by this disaster, the government set about preventing any repetition by ensuring that sufficient rice, wheat and other cereals would always be available to feed the people, at least minimally. Growth of other crops and maintenance of environmental amenities took second place.

Once the grain base was reasonably secure, though, the Chinese were able to diversify their food base, which they have done by encouraging limited free enterprise – people are allowed to raise vegetables, chickens, pigs and various other crops for personal profit. And agricultural communes have increased their productivity through experimentation with crops and farming methods, even carrying on selective breeding programmes to develop new, higher-yielding varieties of crops, which they often exchange with other communes.

The Chinese have become leaders in developing ecologically sound farming methods, including the extensive use of natural fertilisers and non-chemical methods of controlling pests. It was the Chinese, for instance, who developed a way of fertilising rice paddies by growing a water fern with a mutualistic nitrogen-fixing bacterium. The method has now been widely adopted in Asia and is contributing to the region's increased rice production.

Recognition is growing in other poor countries, especially in the tropics, that modern agriculture, as practised in the temperate zone nations of North America and Europe, is ill-suited to tropical climates, soils and ecosystems. Some experimental research to develop systems that work and that might be indefinitely productive in tropical settings has been carried out with promising results, but much more is needed.

In addition, more and more people in poor nations are beginning to understand the need to preserve large areas of relatively undisturbed land on which natural ecosystems can continue to function and some measure of biotic diversity can be preserved for the future. Such reserves can help to maintain the essential services provided by ecosystems; they also serve as sources of 'spare parts' for systems in areas more heavily impacted by the activities of *Homo sapiens*. Fortunately, the setting aside of reserves has begun in some of the most threatened regions – in the humid tropics, where the greatest diversity of life is found and natural ecosystems are now swiftly vanishing. Brazil, for instance, plans to set aside some 600,000 square miles (almost a fifth of its area) in well-sited reserves and parks; Costa Rica plans to expand its parks until they occupy a tenth of its territory.

By way of contrast, barely 3 per cent of the United States land area has been set aside for national parks and monuments, although another 6 per cent is in protected wilderness areas and wildlife reserves, and a hefty proportion of the protected land is in just one state – Alaska. The proportion of Britain's land ostensibly set aside for nature is an impressive 40 per cent, but most of that is a long way from undisturbed or being protected in any meaningful sense. Much is still being farmed or is open for development; a large portion belongs to the Ministry of Defence. If the degree of protection on most of this land is a scandal, what happens to the unprotected remainder is worse; even semi-natural bits such as hedges were until recently being ploughed under with the government's blessing.

Unfortunately, the plans of many developing nations for establishing national parks and preserves exist only on paper; whether the resources to implement them can be made available, and whether the conservationist elements in those societies can defend the reserves from exploiting elements, remains to be seen. The Costa Rican park administrators have insufficient funds for proper protection of the reserves already established, let alone new areas. In Brazil, the balance of power in government agencies still rests with the exploiters. But at least the needs of conservation have been recognised on paper, and that is a step forward.

217

134. **A wetland reserve in Britain.** The Bure Marshes National Nature Reserve protects a fragment of remaining fen and carr woodland in the Norfolk Broads. Wetlands are often unusually productive ecosystems. Much of Britain was originally marshland, but most of it was drained long ago and turned to agriculture. This protected area is the home of the rare swallowtail butterfly shown on p.133.

All the protection and good intentions in the world, however, cannot save what remains of Earth's once-rich flora and fauna if humanity continues to expand not only its numbers but its plundering of the planet's non-renewable mineral resources to support wasteful, environmentally malign lifestyles. By burning its capital in this way, humanity is mortgaging its own future as well as endangering the lives of its fellow spaceship passengers. It is crystal clear that no physical quantity can grow forever on a finite Earth, and that humanity is rapidly dissipating its inheritance of non-renewable resources. There are indeed 'limits to growth'. And, even in historical terms, those limits are appallingly close.

That means the era of what US economist Kenneth Boulding called the 'cowboy economy' must come to an end. A cowboy economy is one with a very high level of 'throughput' — it specialises in turning natural resources into rubbish as fast as possible. It is a reckless, exploitative economic system, based on two demonstrably false premises: that Earth has infinite resources, and that the ecosphere has an infinite capacity to absorb abuse. These premises underlie most of what is taught in university courses in economics, in which growth of the conventional economic measure of throughput, the gross national product (GNP), is considered an unalloyed good.

Among the brightest developments in recent decades has been the emergence of the idea of a steady-state economy and the related notion of a 'sustainable society' — one that does not depend for its health on perpetual growth. Such an economy is sometimes called a 'spaceman economy' because in space vehicles all resources are strictly limited and must be carefully used and recycled. Planets are just large space vehicles; if civilisation is to have a long future, the spaceman economy must replace the cowboy economy.

Over the long term, Earth could support a relatively small human population, with each person living quite extravagantly. Or it could support a much larger number of people living near the subsistence level. Exactly how many people could be permanently maintained cannot now be calculated, since it would depend to a large degree on the character of future technologies and socio-political systems. It would also depend on whether, how much, and how soon Earth's ecosphere could recover from the damage inflicted on it during the current human population outbreak. At least for the foreseeable future, though, the number is clearly far below today's five billion.

In a properly designed steady-state economy, the *product* of human numbers and some measure of an average individual's impact on the environmental systems of Earth would be adjusted to a level that could be sustained over the long term. An essential element would be to reduce human dependence on non-renewable resources and increase

219

reliance on renewable ones, but without allowing the productivity of the renewables to be undermined (as occurs when aquifers are prevented from being recharged or soil is allowed to erode). A key aspect of a steady-state economy is that it would sharply reduce the throughput of materials used by a society. Instead, metals and other non-renewable materials would be almost entirely recycled. Similarly, the system would be designed to minimise wastage of energy. Energy systems would be heavily based on solar income, with fossil fuels conserved as much as possible and used only when substitutes were unavailable.

Describing a steady-state economy is much easier than establishing one, however. Major problems of converting to a steady-state economy include finding ways of gradually reducing material throughput and encouraging modernised societies to make the transition from living on inherited capital to living on income. Herman Daly of Louisiana State University is a leader among a handful of economists who have grasped the basic physical and biological laws that place constraints on human activities; he has tried to design an economic system to operate within those constraints.

Daly has proposed that societies establish depletion quotas for non-renewable natural resources. Each government would place limits on the total amount of each resource that could be extracted or imported into its nation each year and allocate the rights of extraction among potential exploiters. The allocation could be accomplished in any of several ways; in capitalist nations, extraction rights might be sold at auction.

If depletion quotas were more or less universally established, the rates of exploitation of non-renewable resources would quickly level off, and recycling of materials would be greatly encouraged as the price of each resource was driven up. Many substances that now go into junk yards, sanitary land fills and toxic waste dumps, or up chimneys, out of exhaust pipes and down sewers would become too valuable to discard. Instead, recovering and re-using them would be economically rewarding. Depletion quotas on fossil fuels and uranium supplies would give further stimulus both to conservation of energy and to research and development of renewable energy sources.

Best of all, the human assault on Earth's environmental systems would be greatly alleviated. Impacts associated with the severance of mineral resources – mining, refining and transporting – would at least cease growing. Impacts connected with the effluents of resource use should decline as well, independent of any advances in pollution control. The numbers and weight of vehicles would be reduced; high-speed, lighter-weight trains might well replace relatively energy-inefficient aircraft for short and medium length routes; and improved mass transit systems in urban areas would become economically attractive. Then, by gradually reducing depletion quotas as rates of throughput diminished, governments could wean their societies away from dependence on non-renewable mineral resources while stretching supplies to sustain civilisation during the transition to a full steady state.

It may sound simple, but of course there would be enormous problems in managing the changeover. Besides developing the political and social will, there would be vast difficulties in dealing equitably between (as well as within) nations – and, for instance, finding ways to prevent the rich from taking over, in one way or another, the quotas of the poor. Indeed, early in the transition, poor nations should be given preferential access to what remains of Earth's accessible mineral riches – higher depletion quotas, not only to compensate for still expanding populations, but sufficient to raise living standards to reasonable levels. Daly and others have considered at some length the kinds of mechanisms for redistribution of wealth and power that would be required to manage the transition and achieve a sustainable society.

There would also be special difficulties in protecting three critical, nominally renewable resources. Soil and groundwater in many areas are now being depleted at rates that make them, in essence, non-renewable, but in theory at least they could be husbanded and managed so as to move them from the class of 'capital' back to 'income'. Agricultural systems have been developed that minimise soil losses on cropland, and economic incentives can be used to encourage farmers to adopt appropriate methods. The same can be said of grazing land that has suffered abuse and degradation. It is possible to restore such land to productivity if the process has not advanced too far. Unfortunately, at present, population pressures and poverty often combine to prevent restorative measures until deterioration has become essentially irreversible.

The third crucial renewable resource, the millions of species and billions of populations of Earth's other organisms, cannot be renewed on a time scale of significance to human beings. If that irreplaceable resource is allowed to be substantially depleted, even under the most favourable conditions for the mechanisms of speciation to function, millions of years would be required to restore the diversity of the biosphere to anything resembling its former glory.

Of course, that renewed diversity would be very different in detail from that existing today, reflecting the loss of many major groups of organisms. And the length of time needed for restoration of the ecosphere would depend in part on the array of organisms that survived the extinction episode. It would also depend in large part on the conditions that prevailed afterwards, such as climates, soils and the

221

presence or absence of ecologically disruptive creatures such as *Homo sapiens*.

The entire notion of human society gathering its collective wits about it and cooperatively moving towards a sustainable society may seem preposterous in a world that today appears hell-bent on moving in the opposite direction. But remember that the thought of reaching replacement reproduction in the United States before the year 2000 seemed preposterous in 1968, and yet a mere five years later it had happened. When the time is ripe, human societies can change with extraordinary swiftness. With appropriate motivation, considerable economic restructuring can also occur in a surprisingly short time, as Britain and the United States both discovered moving into and out of wartime economic systems at the beginning and end of World War II. Indeed, the economic recovery of Germany and Japan after that war show what *can* be done with sufficient will, planning and international cooperation.

The world is already moving in the right direction in many ways. In reaction to the oil crisis of 1973–4, there was an upsurge in energy conservation, notably in the traditionally profligate United States. Many industries have taken up cogeneration – using waste industrial process heat to generate electricity – a system long used in Europe but ignored in the United States until higher energy costs forced the issue. American cars have shrunk in size and weight; homes, offices, shops and factories have been insulated; and appliances have been redesigned for greater efficiency.

As a result, per capita energy use in the United States was about 5 per cent lower in 1984 than it was in 1975, and energy use per dollar of gross national product (adjusted for inflation) was more than 20 per cent lower. Before 1975, total commercial energy use in the United States had been increasing by over 4 per cent a year. Virtually all the decline in energy use has come from a reduction in the nation's petroleum consumption since 1979 of about 16 per cent. Similar declines in oil consumption have occurred in Japan and parts of Europe.

The oil glut of the mid-1980s, which caused a spectacular drop in oil prices, was in part a result of conservation, in that global capacity to produce oil had temporarily outstripped demand and oil-producing countries were competing fiercely to sell their oil at the expense of others. As the overproduction drove down prices, the economy of countries dependent on oil income became less stable, even though the lower prices were in other ways economically beneficial. In any event, the glut was bound to be short-lived; sooner or later, the realities of resource depletion will catch up with the market. Those in the know predict the return of oil shortages by the 1990s. Abandoning conservation measures would be folly of the first magnitude, even if oil production kept supplies abundant for a few more decades.

In less than half a century since World War II, a very short time as history is normally measured, many revolutionary trends in the direction of a sustainable society can be detected. For instance, in most Western societies a rather effective war against racism and religious intolerance has been fought, and many battles won. Britain and other European nations no longer hold sway over masses of people in exploited colonies. Legal segregation of races no longer exists in the United States. Many nations are pressuring South Africa to modify its loathsome policy of apartheid. And Britain, after centuries of repressive policies, is at last seeking a reasonable solution to its Irish question (albeit in the face of massive prejudice on both sides).

Anti-semitism remains alive and well in the Soviet Union, Arab nations and (at a lower level) much of the rest of the world. But a Jewish state does exist, with the support of many other nations, and overt anti-semitism is sufficiently disapproved of that even those at war with Israel usually design their propaganda to portray political rather than religious enmity. And in Israel, a substantial minority has objected to the racist treatment of Arabs within that state, and a majority tries to counter the religious and racial bigotry of the ultra-orthodox.

The problem of sexual discrimination is similarly improving on the whole. In many areas, at least in Western countries, the condition of women has been substantially improved in the past fifty years, although even in those nations there is still a very long way to go. Women have been heads of state in Great Britain, India, Sri Lanka and Israel, and they hold numerous lesser political offices throughout the world. In most businesses, the notion of women holding top executive positions is no longer novel.

In science, long a stronghold of sexism, barriers are falling. In our own department of biology, which for a long time was all but exclusively male, some of the brightest and most influential young stars are women. Women are demanding, and in some cases gaining, the right to be priests and ministers in religions that once barred them from significant roles. It is difficult to believe, given the changes elsewhere, that the Roman Catholic Church will much longer be able to deny women the right to be priests.

Unfortunately, progress for women seems to be at a near standstill in the Soviet Union, despite much lip service paid to equality and almost universal female employment. And the conservative tide in the United States has also set back the cause of women's rights – especially by the defeat of the equal rights amendment and the repeated attempts to repeal rights to abortion and restrict access to birth control.

223

Sadly, in many developing nations, especially in Africa and some Muslim societies, women still have pathetically few rights except to work, marry, bear children and die. It is noteworthy that developing countries that have elevated the status of women and granted them equal rights often have been comparatively successful in reducing fertility and increasing the well-being of their citizens. China is an obvious example, but there are many in the so-called free world as well: for instance, Taiwan, South Korea, Sri Lanka, Costa Rica, Tunisia – and of course, Hong Kong and Singapore.

The critical point is that things are changing, and much of the change is in a direction that, all else being equal, could lead to a sustainable social system. An important characteristic of such a system must be that it be sufficiently equitable that almost all members of the society would be willing to cooperate and sacrifice for a common good. There is no reason whatever for the have-nots and the oppressed of the world today to help the haves and the oppressors to save *their* comfortable world. The issue of creating a sustainable society cannot be separated from the issue of social justice.

The rate of increase in justice in the world seems depressingly slow, indeed glacially slow, especially in places like the Soviet Union (where an optimist can none the less detect them). But with a little historical perspective, they appear more rapid. After all, less than fifty years ago, segregation by race was legal in the United States, and the British ruled by force an empire of dark-skinned people. Many Americans alive today were born when women could not vote, lynchings of blacks were commonplace and Jews were effectively banned from many schools, clubs and professions. Many Russians can remember when Stalin was exterminating millions of peasants as a matter of policy; many Europeans still living were adults during the era of Adolf Hitler. In the lifetime of the grandparents of those people, it was legal to buy, sell and hold slaves in the United States; and a few generations before that, the King of France ruled by divine right.

No sensible person would contend that the trail away from such horrors is irreversible. The fate of Cambodia, if nothing else, has put the lie to that – as has Soviet behaviour in Afghanistan, American behaviour in Vietnam and Central America, and Britain's reaction to the Argentine occupation of the Falklands. And one must never forget that both East and West have financed the development and maintenance of weapons systems designed to murder hundreds of millions of innocent people.

But there is still a chance that humanity will turn away from the war system and that those weapons will never be used. The possibility of abolishing war has been high on the human agenda through much of this century, and the acquired capacity for collective suicide –

135. **Russian military equipment in Afghanistan.** Failing to learn from the American debacle in Vietnam, the Soviets have attempted to impose by force on the Afghan people a regime friendly to themselves. Their adventure may prove as costly in economic terms and in lives as the American one. Too many nations large and small are behaving as if the world had not changed since the Congress of Vienna in 1815; and the price of that mistake could be the demise of civilisation.

humanicide, if you like – will surely keep it there. And even though a majority of human beings lack what most of us would consider minimal political freedom, perhaps two billion people do have it – as many people as lived on Earth when we were born.

That is progress. It has happened fast in historic terms, and the tools are at hand to accelerate the pace. Global communications have made human rights an international issue in a way it has never been before. Even governments like those of the Soviet Union, Vietnam and South Africa are not immune to the pressures of international public opinion; and to one degree or another, their people are aware of the freedom of others. Indeed, the trends of the last few years are truly heartening, as one after another apparently entrenched repressive regime has toppled, from the generals in Argentina and Samoza in Nicaragua to Baby Doc in Haiti and Marcos in the Philippines.

We do not wish to minimise the grim social trends treated earlier – increasing ethnic strife and terrorism, a yawning gap between rich and poor, the resurgence of fundamentalism and reactionary know-nothingism in the West, the obdurate xenophobia of the Soviet Union. But it is important to realise that a similarly grim litany could have been assembled for virtually any historic era since the rise of civilisation. And we would claim that the hopeful trends, taken together, are unique to today's world.

There is even a little good news on that grimmest of topics – nuclear war. The bellicosity of the Reagan administration and Soviet behaviour in Afghanistan created an atmosphere of fear in the early 1980s that

225

spurred renewed efforts by concerned people around the globe to seek ways of avoiding a thermonuclear Armageddon. The concept of a freeze on the deployment of nuclear weapons became so popular that the United States House of Representatives nearly passed a resolution favouring it. Both Houses of Congress have passed, by large majorities, resolutions calling for renewed negotiations among the nuclear powers of a comprehensive test ban treaty – a long-awaited companion to the Partial Test Ban Treaty signed by the Soviet Union, the United States and the United Kingdom in 1963.

Action by citizen groups has been extremely important in marshalling public opinion towards ending the arms race. Physicians in the United States, the Soviet Union and other countries, realising that they would be unable to give significant assistance to the massive numbers of injured people in a large-scale nuclear conflict, have organised themselves and their colleagues into a worldwide pressure group for peace. Members of several other professions have followed suit.

136. **An international meeting of scientists.** Among the most effective ways to increase understanding between competitive nations are meetings and exchanges of information and ideas among private groups, such as these scientists meeting at the International Institute for Applied Systems Analysis (IIASA) at Laxenburg, Austria. Speaking here at a recent meeting is Vitali Kaftanov of the Soviet Union. On the left in the first row is Ted Munn of the United States. Others in the audience are from various other nations on both sides of the Iron Curtain.

Many individuals in the United States military services also have become disturbed about the extent to which American forces are nuclearised. Numerous retired military leaders, including Admirals Gene La Rocque, Eugene Carroll and John Marshall Lee, General William Fairbourn and Colonel James Donovan, have become prominent critics of United States (and Soviet) nuclear policies.

It is important to remember that well-trained military people are familiar with issues of military ethics – including the question of how much collateral damage to civilians is permissible in order to achieve a military goal. That issue vexed civilian and military leadership in

Britain during World War II. The decision in 1942 to initiate terror bombing of German cities, an action directed solely against civilians, was highly controversial. It was justified in the minds of Churchill and others on the bases of both extreme emergency and the nature of the Nazi regime. But in the minds of many (apparently including numerous British civilians who suffered under German bombing), it remained a highly questionable decision.

The final judgement was rendered after the war. In Westminster Abbey, a plaque was installed listing by name the men of Fighter Command who had given their lives to win the Battle of Britain. No such honour was bestowed on the men of Bomber Command, and its chief, Air Marshall Arthur ('Bomber') Harris, was snubbed after the war. He was not given a peerage, as were other prominent leaders, and he finally retired to his native Rhodesia. Yet the young air crews of Bomber Command, if anything, showed even more bravery on the long night-time missions they were ordered to fly over Europe than did fighter pilots on short day-time sorties.

The point is that, even in times of great national crisis, military men and political leaders can take moral positions. At the peak of the Nazi blitz, when the very survival of England was at stake, many British military officers argued that bombing attacks on Germany should be limited to military targets and terror bombing eschewed. Similar arguments went on in the American armed forces during the Vietnam conflict over such issues as the morality of free-fire zones.

Small wonder that many in the United States military are uncomfortable with a strategic posture that assures that many tens of millions of enemy civilians would perish in an all-out war. They are not for a moment fooled by statements from either Soviet or American propagandists that weapons are aimed only at military targets. Factories, airfields, seaports, railroad yards and various government offices are all military targets – and they are almost always located in or adjacent to populated areas. It would be small consolation to Londoners that the bomb was aimed at Heathrow, to northern Virginians that the target was the Pentagon, or to the people of the San Francisco Bay area that the targets were the Alameda Navy Base or Silicon Valley. And vaporised Moscovites would not care that NATO warheads were aimed at the Red Army headquarters.

In short, intelligent military men are increasingly distressed at the intrinsically indiscriminate nature of any large-scale nuclear war and increasingly inclined to argue against extreme dependence on nuclear weapons to achieve national security. This attitude is certain to have been sharpened by the results of recent studies of the possible effects of nuclear war on the biosphere. Now, more than ever before, it is apparent that not only would vast numbers of civilians – including

227

children – in combatant nations be massacred in World War Three, but tens of millions in non-belligerent nations are also likely to be killed.

We believe that rising concern among the military about the course of the arms race, at least in the West and possibly also in the East, could be one of the most heartening trends as the twentieth century draws to a close. To turn an old saying on its head, war is too serious a business to be left to the likes of Ronald Reagan – to people whose knowledge of deadly conflict comes from wartime propaganda films

137. **The Moscow link.** As part of the conference that first publicly unveiled the scientific findings on nuclear winter, an exchange of views by closed-circuit television between Western and Soviet scientists took place. Shown on the giant screen in front of the audience in Washington is Dr Yuri Israel, a member of the Soviet Academy of Sciences, who participated from Moscow. The wide scientific recognition that a large-scale nuclear war will produce catastrophic environmental consequences has not yet caused a perceptible change in the policies of nuclear powers.

and who have never heard a shot fired in battle.

The recent discovery that a nuclear war might be followed by a nuclear winter has led to another positive trend. People in non-aligned nations now see their fates even more strongly intertwined with those of the superpowers, and some have again begun pressing for nuclear disarmament. Shortly before her assassination, Indira Gandhi visited the United States for the 'Five Continents Peace Initiative' – and mentioned nuclear winter as an important source of concern. Pressure from outside the superpowers was a key to the establishment of the Partial Test Ban Treaty, which prohibited the detonation of nuclear weapons in the atmosphere and has doubtless saved humanity from an epidemic of cancers and birth defects. Olaf Palme's last official act as prime minister of Sweden before his assassination in 1986 was to sign another Five Continents appeal to the superpowers to conclude a comprehensive test ban treaty.

Even aligned nations are beginning to show real signs of wanting to break the deadly spiralling of the arms race. Peace movements have become very powerful in western Europe; New Zealand has refused to allow United States ships with nuclear arms to use her ports, and Australia has debated similar action. A day that President Eisenhower long ago foresaw may yet arrive: when the people of Earth want peace so badly that their leaders actually let them have it.

CHAPTER TEN

It's not the Planet Mongo

In spite of Earth's deteriorating condition, many positive trends can be discerned, as we have seen. And there is one very powerful reason to be optimistic about the future of our spaceship and its precious living cargo: Earth is not threatened by something like a runaway planet bearing relentlessly down upon it. Rather, its peril is entirely traceable to the behaviour of a single species, *Homo sapiens*. It's not a runaway planet Mongo coming to destroy us – in the now time-worn observation of the famous cartoon character Pogo, 'We have met the enemy and they are us.'

The end of Earth as we know it is desired by none of the five billion people now on the planet, but they must work together to avoid it. That end is the collapse of civilisation, accompanied by enormous damage to Earth's ecosphere. It could come about gradually over the next half-century as a consequence of a continuation of both human population growth and behaviour patterns that degrade the planet's ecosystems. Or it could occur with lightning swiftness any day, by design or accident, if an appreciable portion of humanity's stock of nuclear weapons were detonated. One might crudely characterise these different potential endings to our world as the whimper and the bang.

The whimper could be avoided by steps that are rather simple in outline, though quite difficult in practice. They are steps that humanity nevertheless in large part *already knows how to take*. The growth of the human population must be halted and a slow decline initiated towards a size that could be sustained over the long run. As we've already indicated, that size would be determined, among other things, not only by the lifestyle lived by the members of that eventual population, but by the lifestyle lived by all people between now and then. The reason, of course, is that the number of people Earth will be able to support in the distant future will depend on how much of humanity's irreplaceable inheritance is squandered in the meantime – and especially on the scale of damage done to the systems that provide society with renewable resources.

229

138. **Choices.** These soldiers are lined up in a trench to kill their countrymen in Afghanistan.

139. **Choices.** These people have lined up at a trench to improve the environment – by planting trees in Sheshemane, Ethiopia.

Numerous steps could be taken immediately to safeguard that inheritance. Some are already underway – China's one-child family programme with its goal of *reducing* its population size is perhaps the single most important step yet taken. Others include attempts to reduce environmental impacts, to conserve non-renewable resources and develop ecologically sound, sustainable agricultural technology. The latter goal, unfortunately, has yet to gain much attention from development agencies and governments. Universal adoption of the World Conservation Strategy would be another important advance. Still, the task of speedily gaining control over population growth worldwide remains humanity's most daunting challenge.

The majority of the nations' leaders have yet to recognise that the world is already overpopulated and that humanity's objective must be a smaller, indefinitely sustainable population. That no one at the moment knows what the appropriate size might be does not matter. Everyone knows the *direction* in which it lies, and decades will be required even to halt growth in most nations. The optimal size to which the population of each nation should shrink thereafter can be the subject of investigation and debate over a long time.

The challenge now is to get moving as fast as possible. Even a small delay in reducing births in a growing population means a much larger population size at its peak; even a small early reduction in the growth rate brings huge dividends later. This is one reason that political leaders have so much difficulty coming to grips with population problems: the results of small changes in rates may be substantial, but the pay-off comes much later. Moreover, with prevention, the disaster that motivates the behavioural change never appears – and so has much less psychological reality than the disaster that does occur and must be repaired. Thus, while an ounce of prevention is indeed worth a pound of cure, it is seldom as compelling.

While the world's leaders have been slow to recognise the true dimensions of the threat posed by overpopulation, when recognition has come, one of the first things to be sacrificed has often been individual freedom. Freedom certainly suffers from the disease of overpopulation as votes are diluted and sheer numbers of people restrict the freedom that individuals might otherwise enjoy. But freedom suffers when cures are attempted as well. In 1976, Indira Gandhi's government used coercion to push its programme of sterilisation in India and was voted out of office in 1977 as a result.

It is not difficult to imagine what the response might be in various parts of the world if governments suddenly realised, too late, that the very survival of their nations, to say nothing of civilisation, depended upon reversing the expansion of their populations. The potential for extreme restriction of freedom, to say nothing of atrocities, in

231

establishing population control, is present even in the relatively few nations that today can be considered genuine democracies. The end of civilisation might well be heralded by a period of unprecedented repression as societies thrashed around trying to do quickly and by force what should have been done gently through persuasion and education over the last forty years.

What then is the ethical course for solving the population problem before time does run out? Is government control of fertility justified now? Our answer is that in some form it is not only justified but morally required worldwide. The size of a nation's population is surely as legitimate a concern of its government as is national security. Indeed, the two are closely inter-related; and beyond a certain minimum, especially in the modern era, numbers do not confer military security.

Governments normally have the power to regulate the number of spouses an individual can have; and, as India, China and a few other nations have demonstrated, governments can both justify and exert power to regulate family size. Since the consequences of continued human population growth would bring about the end of nation states, the collapse of civilisation and the premature deaths of billions of people, it is difficult indeed to imagine a more appropriate realm for governmental concern and action than regulation of family size. The most difficult ethical question thus becomes not *whether* governments should now attempt to control population size, but *how* they should.

Because of the high per capita use of resources and environmental impact of the rich nations, their overpopulation is more serious for Earth *as a whole* than is overpopulation in poor nations. Clearly, in the United States, Canada, the United Kingdom, Japan and most other overdeveloped countries, the first control measures to be tried should be non-coercive ones designed to remove incentives for having large families, to ensure full access to all safe and effective means of birth control, and to put the moral (and educational) power of the government firmly behind the notion of 'stopping at two'. Given the present demographic situations in these post-industrial nations, we suspect such action would be enough to end growth relatively soon in those that still have expanding populations and begin the required slow decline.

Unfortunately, such simple methods are not likely to be sufficient for most developing countries. Each nation will require programmes designed by and for the local people. As mentioned earlier, the greatest successes in reducing birth rates have occurred in countries that have made progress in providing education, especially of females, and improved health programmes which reduce infant mortality and increase life expectancy, as well as strong family planning programmes.

140. **Choices.** Instead of going to school, this child from a less developed country begins her day with a long, difficult trek to fetch water from an unsanitary source.

141. **Choices.** A relatively simple and inexpensive change can make an enormous difference for the lives and health of poor people. This Pakistani child is drawing clean water from a convenient community stand pipe.

Thus the social changes that set the stage for family limitation are all appropriate social goals in their own right. So not only does controlling population growth by itself benefit people in developing countries, but the types of development that are most immediately beneficial – and that will contribute to further development – also foster population control.

Other policies that have sometimes helped lower birth rates include provision of food supplements and social security programmes. In some countries, government-sponsored peer pressure and economic incentives and disincentives such as eligibilities for housing and education have been successfully applied.

Many of the measures that seem likely to reduce birth rates in poor countries also enhance human freedom. Improving the condition of women is an outstanding example. In many African nations, women both do most of the farm work and also have larger families than they would like, because fertility is considered the principal sign of manliness. Young girls in a number of tribes are by tradition coerced into being circumcised, often with adverse consequences for their health and marital relations.

Granting women more freedom to determine their own destiny would almost certainly lead to a dramatic decline in birth rates. Liberating women without destroying the fabric of African societies is clearly a task that will require enormous diplomacy, compassion and patience. Yet that fabric is already being rent by the stresses of deepening poverty, failing agricultural productivity and family separation (because men commonly migrate to cities seeking jobs while women remain behind to keep the farm going).

Many freedoms could be gained or preserved in poor countries if population growth were halted and reversed, even though, of course, the freedom to reproduce would be curtailed. The growing possibility of a global disaster justifies extraordinary action now to embark on relatively humane and non-coercive paths to population control. Some freedom of reproduction must be sacrificed to preserve or secure other freedoms, such as freedom from hunger, want and fear. Delay simply increases the chance that draconian measures will eventually be required, and that even they will be applied too late. Individuals who oppose mild and humane restrictions on reproduction now are encouraging an enormous further loss of both human freedom and human lives in the future.

No one would pretend that stopping population growth in a poor country is easy; there are many, seemingly more urgent claims on limited resources, and too few political leaders are as yet committed to a serious effort. But how to do it is no longer a mystery. Modern contraceptive methods may not be perfect, but some method is

142. **Choices.** This beach at Portsall, Brittany, was covered by a huge oil spill from the tanker *Amoco Cadiz* in 1978 which marred its beauty, threatened local fisheries and wreaked a great deal of damage. A massive clean-up operation restored the coastline and the shore birds have now returned, but oceanic pollution is gradually increasing, and such accidents are, to a degree, an unavoidable cost of the large-scale extraction and use of fossil fuels.

143. **Choices.** Kurumba beach in the Maldive Islands near the southern tip of India remains, for the moment, pristine.

adequate for most people most of the time. Abortion, undesirable though it may be in many ways, is available as a back-up (where legal), and sterilisation can prevent extra births after families are complete.

Another example of a step in the direction of a sustainable world is the trend towards conservation of energy in the West. Conservation, of course, reduces the pressures on non-renewable fossil fuel and uranium resources, and thus at least slows the rate of capital-burning by humanity – especially by the groups of people most heavily engaged in that destructive behaviour, the industrialised nations. Beyond that, however, the use of energy is directly or indirectly involved in almost all environmental deterioration.

This is true whether the energy in question is obtained by the burning of fuelwood in a simple stove in a village hut or the burning of coal or oil in a modern power plant. Wood burning causes deterioration through deforestation or destruction of woodlands, which leads to soil erosion and desertification. Extracting fossil fuels is also destructive of the land, most massively in the case of strip-mining for coal. Processing and transporting fossil fuels also have deleterious environmental effects. And the burning of either wood or fossil fuels inevitably releases a variety of more or less noxious combustion products into the environment.

Even nuclear fission has environmental impacts. The mining of uranium creates serious local problems of disposal of radioactive and in other respects highly toxic tailings. Then there are the problems of low-level radiation at the power plant, the discharge of heat and disposal of the highly radioactive wastes after use. Finally, there is the ever-present danger of a catastrophically large accidental release of radioactivity from the plant, a danger of which everyone was rudely reminded by the event at Chernobyl. Indeed, several aspects of that accident turned out to be problems that the public in the West had been assured by the nuclear power industry could never happen.

A troublesome problem now looming in some industrial nations is that of decommissioning nuclear power plants that have reached the end of their useful lives. In the next few decades, dozens of plants in Europe and North America will be ready to retire. Decommissioning them is likely to prove a very expensive and hazardous business, especially since no back yards have been offered for the permanent storage of the highly radioactive components.

Energy conservation and lowered demand for electricity of course reduces the need for power plants of all kinds. With conservation, and as the process of planning, siting, building and commissioning safer plants has become increasingly time-consuming and expensive, plans for constructing new nuclear power plants have been substantially scaled down. The result is a diminution of both the potential environmental impacts of nuclear power and the chances of a disastrous accident.

Similarly, when energy conservation results in the burning of less fuel in oil, gas or coal-fired power plants, emissions of combustion products are accordingly reduced. Building smaller cars leads to less environmental disruption from the mining of metals, less manufacturing of plastic parts (and disposal of toxic by-products) and less destruction of ecosystems through the paving of roads and parking lots. To the extent that energy needs can be met by renewable sources such as solar (active or passive) and wind, the environmental effects of the traditional sources are further reduced. And because the renewable sources are decentralised, their own environmental effects may be more equitably distributed as well.

But energy conservation, by itself, cannot adequately lighten the environmental burden. This is seen clearly in the urgent need to abate acid rain, which threatens to destroy ecosystems over large sections of the Northern Hemisphere, cutting off both the delivery of vital services from ecosystems and the supply of goods (forest products, fish, crops) for society. To reduce this threat requires, among other things, installing pollution control devices on the exhaust stacks of hundreds of power-generating and industrial plants. But such devices can be costly, and many of the most polluting plants are in areas and represent industries that are already threatened economically. Thus the long-term benefits to society as a whole of controlling the pollution must be balanced against the short-term costs to stockholders who may lose profits and to workers who may lose their jobs.

This choice is not a new one, though, and in most Western societies it has been made, at least in principle: polluters have been forced to internalise the costs formerly imposed on the public at large. To the degree feasible, industries must prevent the emission of pollutants; where this is too costly, or it is too late (as in the case of toxic wastes dumped decades earlier), the offending companies must pay for cleaning up.

The health consequences of ordinary air pollution are well documented. Even so, it took several decades before the case was strong enough to justify establishing tough nationwide air pollution regulations in the United States, and those regulations are still in several respects inadequate.

Acid rain presents a more complex problem, however. It can have detrimental effects on human health, especially through acidified drinking water supplies, but it mainly attacks other organisms – indeed, entire ecosystems. The attacks are subtle, varied and complex in their manifestations and usually slow to develop. Depending on the

237

144. **Choices.** Pollution of inland waters is another increasing global problem, in spite of some local successes in preventing or cleaning it up. This effluent was from a felt mill in Blackstone, Massachusetts; it has now been cleaned up following pressure from the Audubon Society.

145. **Choices.** Some fresh water still flows clean and free. This waterfall is in Catoctin Mountain Park, Maryland.

buffering capacity of the system exposed, many years, even decades, may elapse before symptoms even begin to appear. The damage is cumulative, in that the buffering capacity is slowly eroded by the continued acid assault. Perhaps the most insidious aspect of acid rain damage is that, once injury to the organisms is entrained by the collapse of the buffering capacity, it is for practical purposes irreversible on any reasonable time-scale.

It is for this reason that dealing with the acid rain threat is so urgent; much of the Northern Hemisphere has been exposed to increasing acid precipitation for decades, and many areas seem to be on the verge of losing the buffering capacity that has protected them until now. If the level of assault is not lowered very soon, the amount of damage that has already occurred may be dwarfed by the coming epidemic of dying forests and sterile lakes, not to mention the possibility of significantly reduced crop harvests in an overpopulated world.

Another vexing aspect of the acid rain dilemma is that it requires international solutions; pollutants are no respecters of boundaries. If national governments have been slow to come to grips with the subtleties of acid rain damage internally, their record on controlling trans-boundary pollution is even worse. Over a dozen European nations blame each other for their crumbling forests and sterile lakes, while the Reagan administration's refusal to deal with the problem at all has soured relations with the United States' northern neighbour. In 1986, to keep relations from deteriorating to an intolerable level, the administration finally agreed to cooperate with Canada in addressing the problem by funding more research.

But governments have run out of justifications and excuses for stalling. While much uncertainty remains about details of the processes of acid rain damage, enough is known to be more than reasonably sure that abatement of sulphur and nitrogen oxide pollutants would have appreciably beneficial results and would prevent ecological disasters in many areas (it may already be too late for some). And the technology does exist to do that. Even if the costs of installing scrubbers to remove pollutants from smokestacks or using other measures raises the price of goods or electricity, or forces some plants to shut down, such steps must be taken. The costs incurred would be far less than the eventual catastrophic costs society would have to pay for the destruction from acid rain.

While acid precipitation is a problem primarily for the overdeveloped nations that mostly cause it, the build-up of carbon dioxide in the atmosphere is in several respects an even knottier problem for the global community. First, while the contributions from combustion of fossil fuels come predominantly from the rich nations, the developing nations contribute significantly, principally by their destruction of tropical forests. And everyone will pay a share of the costs.

Moreover, unlike acid rain, the carbon dioxide problem is not caused by a trace pollutant like sulphur oxides; it cannot be captured in any practical way from stacks by scrubbers or avoided by burning a different kind of coal. It is an inescapable result of the process of combustion of fossil fuels or wood and of forest clearing. Consequently, the only way to stop producing excess carbon dioxide is to stop burning fossil fuels and wood and stop the wholesale clearing of forests.

It is completely impractical to stop burning fossil fuels entirely, of course, given the human population's dependence on them to support itself – to produce enough food and other crops, manufacture goods, heat homes, cook food, transport people and goods and so on. But the rate of production of carbon dioxide could be significantly lowered, which would greatly alleviate the problem. Simply slowing the rate at which the carbon dioxide accumulated in the atmosphere would allow more time for part of it to be absorbed in the oceans and for both society and biotic communities to adjust to changes in climate.

This slowing of the carbon dioxide build-up process could be accomplished through several measures: by energy conservation, especially of fossil fuels; by switching to other forms of energy (hydro, solar or wind) whenever feasible; and by preferentially using those fossil fuels that produce less carbon dioxide per unit of useful energy derived (oil is better than coal, natural gas is better than either). These measures would have the additional advantages of reducing emissions of various air pollutants, including the precursors of acid rain. A world programme to limit deforestation, especially in the tropics, and encourage regeneration of forests would also help to retard the release of carbon dioxide – as well as preserving the still uncounted values that those forests contain.

Besides carbon dioxide, the build-up of the other trace gas components of the increasing greenhouse effect should be retarded in so far as is possible. Methane production from livestock and agriculture may not be controllable, but the contribution from termites can be curbed by limits on forest clearing. Releases of nitrous oxides from fertilisers and CFCs from refrigerants may not be easily reduced, but society certainly can get along without CFC propellants in spray cans.

The dilemma of greenhouse gases shares with acid rain the characteristic of being a long-term, subtle threat, in which much uncertainty surrounds both the processes and the exact consequences, both for the ecosphere and for humanity. It seems to be very difficult for people to deal with serious problems when the worst possible consequences have yet to appear and there is no way to predict exactly

241

how bad they will be when they do. But once the consequences have manifested themselves, it will be too late to reverse them. And the greenhouse gas build-up in particular is a global problem, caused by people in both rich and poor countries, and can only be successfully addressed through concerted international action. Its causes, moreover, are intimately related to important issues of development in poor countries, which therefore will have to be partners in the effort as well as the disparate, often mutually unfriendly industrial nations.

The environment is not an infinite sink whose abuse carries no penalties, as many societies have learned to their sorrow. Around the globe there are dramatic signals that the capacities of Earth's environmental systems to serve humanity are being stretched to the limit and sometimes broken – signals that are ignored to our great peril. Fortunately, natural systems *do* have considerable ability to resist degradation and to rebound from gross insults, if given half a chance, as the return of salmon to the Thames has demonstrated. But, if civilisation is to find its way towards a future sustainable world, it must avoid as much as possible any further deterioration in those vital life-support systems.

Happily, there are many ways in which societies could reduce the pressure and give ecosystems some chance of recovering. The prime step might be one of the hardest to take, but it is essential: *to permit no development of any more virgin lands*. Since *Homo sapiens* already occupies and uses the vast majority of usable, productive land at some level, precious little is left that has not felt the tread of a human foot. But whatever remaining relatively undisturbed land exists that supports a biotic community of any significance should be set aside and fiercely defended against encroachment.

Plenty of land is available for development that has previously been used in other ways – as farmland, pasture, logged forest or human settlement. All new development should be restricted to such land. Moreover, the most intensive kinds of development (housing, factories, roads, airports etc.) should be sited as much as possible in older, intensively developed areas. Thus new housing developments and shopping centres should replace worn-out old housing and stores, not prime farmland. New motorways should be built over older, lower capacity roads, not allowed to cut new swathes across farms and through forests. Airports should be enlarged if possible, not replaced. When they must be replaced, the old ones could be used for other forms of intensive urban development – factories or shopping centres, for instance – taking advantage of already existing road and power grids and water and sewerage services.

Adopting as a general principle everywhere that development should be planned so as not to degrade the land being used would be a major

146. **Housing being restored.** It is usually far more economical and environmentally benign to restore or replace old housing and other buildings than to open new subdivisions outside a city while leaving the ageing core to decay. In this block of housing, an unrestored residence is flanked by two renovated ones in Islington, North London.

step toward preserving Earth's biotic diversity. Under such a regime, natural forests could not be converted to pastures or farms; fragile or marginal grasslands would not be converted to cropland; and good farmland would not be covered with roads, factories, or housing. Increases in food production would have to come from enhanced yields on already cultivated land – as nearly all of it now does anyway.

Exceptions to the general rule might be made in certain cases, such as areas where new irrigation projects make possible the cultivation of formerly arid land. Such projects, of course, should be carefully evaluated beforehand to ensure that the gains would outweigh the losses and costs, including the biotic ones, in the long term. But compensation for the intensified land use could be made – might even be a requirement – by taking an equivalent tract of marginal land out of crop production and allocating it to a less intensive use, such as pasture.

A corollary principle to the one of not further degrading any land might be to restore as much land as possible to a biotically richer status.

Thus abandoned railways or roads could be turned to open space with encouragement of the growth of native flora and fauna. Hedgerows, windbreaks and strips of natural vegetation should be re-established between fields and farms in the countryside. These can provide small habitats for flora and fauna and corridors for dispersal of plants and animals between larger ones. In addition, they also benefit farms by contributing natural pest control and soil conservation services.

Turning marginal farms into parks, preserves and green belts could be considered a 'higher use' for land, to be encouraged, especially near cities where recreational open space is often scarce. City parks and green belts could be enriched by the planting of native flora, and homeowners could be encouraged to do the same with their gardens. Such improvements would not only enhance people's immediate surroundings and the quality of their lives, it would help preserve the planet's diversity of life by maintaining, if only a tiny bit in each individual instance, a habitable place. And, of course, thousands of tiny instances could add up to a lot of habitability.

All such measures, whether directly addressing the world's serious environmental threats or trying to preserve the remaining ecosystems of the planet, would cost money – no doubt about it. But it would be money invested in the health of both individuals and societies, indeed in the future of civilisation.

One obvious place from which to obtain the resources to start restoring our environment is the swollen military budgets that afflict most of the world's nations. Now is surely the time to do it, for many people have come to realise that war can no longer be, as Karl Von Clausewitz described it, 'nothing more than the continuation of politics by other means'.

Almost everyone now understands that large-scale nuclear war could bring about the end of civilisation; consequently, there must never be such a war. Leaders of both superpowers, and other nuclear-armed nations as well, have acknowledged that a nuclear war could not be won. That being the case, the tens of billions of dollars being poured into nuclear weapons and their delivery systems every year (beyond perhaps a small deterrent arsenal until mutual distrust can be reduced to manageable levels), are simply being wasted on an unusable product.

Yet, even if nuclear weapons were somehow totally banned, large-scale wars with other kinds of weapons would still be unacceptably catastrophic. If nuclear weapons disappeared, the determined militarists would doubtless turn to chemical and biological weapons in an attempt to achieve a similar capability of mass death and injury. If those unconventional weapons were also successfully banned, humanity would soon realise that the destructive capacity of 'conventional' weapons has been vastly increased since World War II. For example, some fuel-air explosives now have yields in the kiloton range — equivalent to small nuclear weapons. So, to avoid unacceptable damage, people who would like to live in a world of nation states that settles its differences through large-scale violence would have to fall back on the weapons used in good old World War II.

But the world today is very different from what it was during World War II. In 1940, Europe was the only net food importing region; today North America is the only major net food exporter. Twice as many people are now alive as were during World War II. Fighting such a war on a much more crowded, much more resource-depleted planet, even with the weapons of the 1940s, would be the greatest catastrophe yet experienced by humanity and might well be a major blow to the integrity of many of Earth's natural ecosystems. And it must always be remembered that, while nuclear weapons might be banned, the knowledge of how to build them cannot be. One might therefore surmise that another World War II-type conflict would quickly lead to a new nuclear arms race and an escalation to a nuclear war.

So nuclear weapons in themselves are not the real problem. The real problem is that NATO and the Warsaw Pact nations find themselves trapped in an escalating, increasingly unstable arms race. If, like almost all arms races in the past, this one leads to war, the chances that the societies on either side would survive in any meaningful way are minuscule, and the future of civilisation itself cannot be assured. Far from contributing to security, accumulating the weaponry and deploying it in a menacing fashion diminishes it in more than one way: by increasing the likelihood of an accidental war that threatens everyone's life and well-being; and by diverting both attention and resources away from the deepening human predicament, which threatens world security far more than minor political differences can and furthers the distrust that helps to fuel the arms race.

The bottom line is that humanity can no longer afford to permit the close connection between nation states and warfare that has existed for the nearly ten millennia since nation states first appeared. Either the connection will be broken, or civilisation will be. 'Realists' may consider world peace an impossible goal, and for all we know they may be correct. But the only realistic course of action today is to strive for that goal.

It seems obvious to us that a bilateral, verified reduction in nuclear arms is a first step that must be taken *now* if humanity is to have a chance for the bright future that is otherwise possible. Certainly, each side will have to take the risk that the other will cheat if such a programme is to be achieved, but with the safeguards now available, *that risk is tiny compared to the risk of an all-out thermonuclear war.*

147. **A greenbelt around San Francisco.** One good way to preserve flora and fauna while making life more pleasant and healthy for city dwellers is to plant or preserve greenbelts around them. The San Francisco Bay Area is home to about five million people, yet a nearly continuous belt of semi-natural parkland and open space surrounds it. Most of this land has been preserved through local citizen action over the past twenty-five years — a period when the Bay Area's population nearly doubled.

148. **A hay meadow near London.** While London cannot exactly be described as having a greenbelt, it does have generous parks within the city and some open space not far outside. This hay meadow, in the London borough of Richmond upon Thames, resplendent with such wild flowers as buttercups and red clover, is only about ten miles from the centre of the city.

The primary responsibility for moving toward a reduction in nuclear arms clearly rests with the United States and the Soviet Union. And, since Western societies are open and the Soviet Union is not, we believe the United States should take the initiative, remembering always that disarmament is not a 'zero sum' game. A gain for one side does not necessarily imply a loss for the other; both players can come out winners.

To those who say that we should not make any move that is not exactly matched by the Soviets, we would reply that if our system is truly superior to theirs, we should hold ourselves to a higher standard. The dispute between West and East is very much about values – the openness of society, the worth and freedom of individuals, and so on. It would be self-defeating for the West to insist on adopting Soviet bloc values, turning itself into a complex of garrison states, in order to 'defend itself' – for it would be giving up the thing it most wishes to defend.

Yet responsibility extends beyond the United States and the Soviet Union, of course. Other major powers of the world, through a combination of example, moral persuasion, and economic action, could bring enormous pressure to bear on the superpowers towards nuclear disarmament. Western European countries, for instance, could choose to strengthen their conventional armed forces while moving towards

247

making their own territories free of nuclear weapons. They might use their trade relationships with the Soviet Union to push that paranoid giant toward reductions in its nuclear and conventional forces.

Britain in particular could eschew its mindless commitment of resources to new Trident submarines and highly accurate, first-strike-capable nuclear weapons. Instead, it could divert those resources to shore up its crumbling economy and invest in its neglected younger generation. The British could set a fine example by being the second major power (Japan being the first) not to waste a huge proportion of its national wealth on armaments.

Other nations could also do much to highlight the similarities between American and Soviet behaviour; to point out that in their murderous and bungled interventions in Vietnam and Afghanistan, for example, each was acting out of national interest and not on any kind of moral principle. It is important that both sides be led to examine closely the exact nature of their differences, stripped as far as possible of the fiendish communist/fiendish capitalist propaganda. When that is done, each might find more common cause with the other than they had thought.

Apart from the potential for generating Armageddon, the gap between Earth's rich and poor nations, basically a North/South gap, is much wider than that between East and West. And the nations of NATO and the Warsaw Pact are all on the same side of that gap. Both blocs could benefit immensely by converting their military competition into a competition to aid the poor. If that is not done, the poor, armed with nuclear weapons (as they soon will be), could become an enormous threat to both. It seems unlikely that three-fourths of the human population will just quietly sink into the abyss while allowing the rich to continue their gluttonous behaviour.

It is entirely within the power of humanity to close the gap between rich and poor and to reduce the human population size to a level at which all people could lead a decent life without degrading the ecosphere. A transition to living primarily on income can be made; agricultural systems can be designed that would be highly productive and would also help support the natural ecosystems in which they are embedded and on which they depend. Societies can turn their backs on racism, sexism, gross economic inequality and, above all, *war* as a mode of problem-solving. And people can learn to value political, social and cultural diversity, and to make the maintenance of organic diversity a quasi-religious duty.

There are, in short, no insuperable barriers to creating a peaceful Earth in which *Homo sapiens* leads a rich existence without overstressing the natural systems that support human life – an Earth on which both biological and cultural evolution can proceed into the

149. **HMS** *Revenge,* one of Britain's Polaris submarines. The United Kingdom's independent nuclear deterrent currently includes four of these nuclear-armed Polaris submarines. During the 1990s, Britain is scheduled to deploy four new Trident submarines, each armed with 16 missiles carrying a total of 128 to 224 nuclear warheads, at a cost of nearly £10 billion for each submarine. If Britain declined to invest in these expensive white elephants, it could save its cash for more pressing social needs and play a leading role in reducing international tensions and building a sustainable future world.

150. *Overleaf:* **Choices.** A thermonuclear explosion symbolises a possible end to our civilisation – a nuclear war followed by a nuclear winter.

151. **Choices.** This sunset over the Serengeti plain in Tanzania symbolises the potential for our civilisation to last thousands of years, powered by a thermonuclear generator sited about 93 million miles away.

249

indefinite future. Unless, of course, the behaviour of our species itself turns out to be such a barrier. We might be cheered by the thought that many present behaviour patterns are quite new in evolutionary terms, the product not of hundreds of millions of years of biological evolution, but of a mere few millennia of cultural evolution. Humanity has the tools in hand to accelerate cultural evolution to the point where patterns that took thousands of years to develop can be altered in decades.

The great hope for civilisation lies in that fact: that people can recognise how the human predicament evolved and what changes need to be made to resolve it. No miracles, no outside intervention and no new inventions are required. Human beings already have the power to preserve the Earth that everyone wants – they simply have to be willing to exercise it.

APPENDIX

If You'd like to know More

We have only been able to sketch an outline of the human predicament in this book, but fortunately you can consult numerous other sources if you wish to be better informed on the topics and issues that we've touched upon. The following annotated bibliography provides both basic documentation for the positions we've taken and access to a wide variety of additional information.

Brown, Lester R., *State of the World, 1986* (W.W. Norton, New York and London, 1986). Latest in an excellent annual series published by the Worldwatch Institute. Deals with a broad range of topics, from economic deficits and decommissioning nuclear power plants to the famines in Africa.

Caldwell, Lynton K., *International Environmental Policy: Emergence and Dimensions* (Duke University Press, Durham, North Carolina, 1984). A useful guide to the history and current structure of international environmental activities.

Cohen, A. and Lee, S. (eds.), *Nuclear Weapons and the Future of Humanity* (Rowman and Allanheld, Totowa, New Jersey, 1986). A fine collection. Be sure to read the outstanding chapter by physicist John P. Holdren, 'The Dynamics of the Nuclear Arms Race: History, Status, Prospects.'

Daly, H.E., *Steady-state Economics* (W.H. Freeman and Co., San Francisco, 1977). Already a

classic; sadly, it has not been read and understood by most of Professor Daly's colleagues in economics.

Ehrlich, P.R., *The Machinery of Nature* (Simon and Schuster, New York, 1986). Describes modern ecology and evolutionary biology, the science behind the environmental predicament.

Ehrlich, P.R. and Ehrlich, A.H., *Extinction: The Causes and Consequences of the Disappearance of Species* (Random House, New York, 1981). How humanity is destroying one of its most precious non-renewable resources and why we should care.

Ehrlich, P.R. and Holdren, J.P. (eds.), *The Cassandra Conference: Resources and the Human Predicament* (Texas A & M University Press, College Station, Texas, 1987). A collection of essays by people who issued early warnings about overpopulation, environmental deterioration, and resource depletion on the situation today.

Ferguson, D. and Ferguson, N., *Sacred Cows at the Public Trough* (Maverick Publications, Bend, Oregon, 1983). Details of how the livestock industry is destroying the American West.

Hardin, Garrett, *Filters Against Folly* (Viking, New York, 1985). The latest collection by the master essayist of the human predicament. You may not agree with all that Professor Hardin says, but he'll make you *think* as few others can.

Harte, John. *Consider a Spherical Cow: A Course in Environmental Problem Solving* (William Kaufmann, Los Altos, Ca., 1985). An instructive venture into ways to analyse environmental problems.

Holdren, J.P. and Rotblat, J., *Strategic Defences: Technological Aspects, Military and Political Applications* (Macmillan, London, 1987). A brilliant analysis by two scientists of the 'strategic defence' idea.

Hughes, J.D., *Ecology in Ancient Civilizations* (University of New

Mexico, Albuquerque, 1975). Best discussion of early environmental attitudes.

Margulis, Lynn, *Early Life* (Science Books International, Boston, 1982). A fascinating summary of the evolutionary history of Earth's first life-forms, as far as was known in 1982. Margulis's ideas on how eukaryotes arose are intriguing and controversial – and may be right.

Myers, Norman, *Gaia: An Atlas of Planet Management* (Anchor/Doubleday & Co., Garden City, New York, 1984 and Pan Books, 1985). A splendid overview of how the planet Earth works and how humanity should take care of it.

Nicholson, Max, *The Environmental Revolution* (Penguin Books, Middlesex, England, 1970). A well-written account of the development of environmental awareness in Britain.

Ornstein, Robert, *Multimind*. A provocative look at the principal tool required to save Earth – the human mind. By an outstanding psychologist.

Population Reference Bureau, *World Population Data Sheet*. An annual update of world population estimates and other useful data. The Bureau also publishes a journal, *Population Bulletin*. (Population Reference Bureau, 777 14th St. NW, Suite 800, Washington, D.C. 20005).

Raven, P.H. and Johnson, G.B., *Biology* (Times Mirror/Mosby, St. Louis, 1986). A superb modern college text – it can provide any intelligent reader with a fine overview of what's happening in the most rapidly advancing science and serve as an excellent reference book as well.

Schneider, Stephen H. and Londer, Randi, *The Coevolution of Climate and Life* (Sierra Club Books, San Francisco, 1984). A superb, comprehensive treatment of how the climate works and how it influences human affairs. A *must* for the library of anyone concerned with the human predicament.

Scoville, Herbert, Jr., *MX: Prescription for Disaster* (MIT Press, Boston, 1981). A keen analysis of the destabilising aspects of the MX system.

United Nations, *1985 Demographic Yearbook*, 1987. Population sizes, compositions, vital rates and so on, for the nations of the world. An invaluable 'neutral' source of data.

United Nations, *1985 Statistical Yearbook*, 1987. Latest annual compendium of data on population, GNP, agricultural production, education, manufacturing, finance and so on.

US Bureau of the Census, *Statistical Abstract of the United States: 1987–1988* (USGPO, Washington, DC, 1987). A goldmine of information on the state of the world's most powerful nation. Contents much like the UN *Statistical Yearbook*, but much more detail on the United States.

Walzer, M., *Just and Unjust Wars: A Moral Argument with Historical Illustrations* (Basic Books, New York, 1977). Contains a comprehensive discussion of military ethics.

Wilson, E.O., *Biophilia* (Harvard University Press, Cambridge, Mass., 1984). A fine, concise volume by an outstanding evolutionist and ecologist arguing that people have a natural empathy for Earth's other living beings.

World Bank, *World Development Report 1986* (Oxford University Press, 1986). An annual volume assessing world development, with emphasis on economic aspects.

World Resources Institute and International Institute for Environment and Development, *World Resources 1986* (Basic Books, New York, 1986). A very detailed assessment of trends in population, environment and resources. Apparently the first of a new annual series.

Index

Figures in italics indicate pages on which illustrations appear

Index

Index